T0245292

Anticandidal Agents

Anticandidal Agents

Awanish Kumar, Ph.D
Assistant Professor
Department of Biotechnology
National Institute of Technology
Raipur, Chhattisgarh, India

Anubhuti Jha
Department of Biotechnology
National Institute of Technology
Raipur, Chhattisgarh, India

AMSTERDAM • BOSTON • HEIDELBERG • LONDON
NEW YORK • OXFORD • PARIS • SAN DIEGO
SAN FRANCISCO • SINGAPORE • SYDNEY • TOKYO

Academic Press is an imprint of Elsevier

Academic Press is an imprint of Elsevier
525 B Street, Suite 1800, San Diego, CA 92101-4495, United States
50 Hampshire Street, 5th Floor, Cambridge, MA 02139, United States
The Boulevard, Langford Lane, Kidlington, Oxford OX5 1GB, United Kingdom
125 London Wall, London EC2Y 5AS, United Kingdom

British Library Cataloguing-in-Publication Data
A catalogue record for this book is available from the British Library

Library of Congress Cataloging-in-Publication Data
A catalog record for this book is available from the Library of Congress

ISBN: 978-0-12-811311-0

For Information on all Academic Press publications
visit our website at https://www.elsevier.com

  Working together
to grow libraries in
developing countries

www.elsevier.com • www.bookaid.org

Publisher: Sara Tenney
Acquisition Editor: Linda Versteeg-buschman
Editorial Project Manager: Fenton Coulthurst
Production Project Manager: Stalin Viswanathan

Typeset by MPS Limited, Chennai, India

CONTENTS

LIST OF FIGURES

LIST OF TABLES

CHAPTER 1

Introduction

Candida is a diploid, polymorphic, opportunistic fungus which can exist in yeast, pseudohyphal and hyphal form and is the most common cause of opportunistic mycoses worldwide [1]. *Candida* are thin-walled, small yeasts (4−6 μm) that can reproduce by budding. There are around 154 species of *Candida*. Of these, *Candida albicans* is the most pathogenic species for man that causes candidosis, followed by *C. tropicalis, C. stellatoidea, C. glabrata, C. krusei, C. parapsilosis, C. guilliermondii, C. viswanathii*, and *C. lusitaniae* in probable descending order of virulence. Importantly, there has been a recent increase in infections due to non-*albicans Candida* such as *C. glabrata* and *C. krusei. C. albicans* causes almost 100% cases of oropharyngeal candidiasis and at least 90% of cases of *Candida* vulvovaginitis [2].

Its growth mode is determined by environmental conditions, i.e., high temperature, high ratio of CO_2 to O_2, serum, nutrient-poor media, etc. [3]. Usually *C. albicans* is endogenous in humans and forms part of the normal commensal flora of the oral cavity, large intestine, and vagina. *C. albicans* is acquired during early postnatal period and remains associated with the human host throughout life. But it becomes pathogenic when the immune functions are impaired resulting in candidiasis which is followed by expansion of the fungus with increased frequency by dissemination via blood stream and colonization of internal organs [4]. This mode of pathogenesis requires the ability to occupy diverse niches of the host and adapt to the changing and adverse conditions present there including that of the host defenses. In order to do so, *C. albicans* has to express a number of phenotypic properties also called virulent factors like yeast-hyphal transition, white-opaque switching, etc. Candidiasis has become the leading fungal infection in the immunosuppressed population. The mortality rates for systemic infection by *Candida* ranges from 50% to 100%. Candidiasis is major cause of morbidity among HIV + patients, intensive care unit patients, and those undergoing cancer chemotherapy or organ transplants. A number of antifungals have been used to treat Candidiasis [5].

Anticandidal Agents. DOI: http://dx.doi.org/10.1016/B978-0-12-811311-0.00001-6

PATHOGENESIS AND VIRULENCE

The capability of *C. albicans* to infect such varied host niches is supported by a wide array of virulence factors and fitness attributes. *C. albicans* is a polymorphic fungus that can grow either as budding yeast, as elongated pseudohyphae, or as true hyphae. Further morphologies include white and opaque cells, formed during switching, and chlamydospores, which are thick-walled spore-like structures. While yeast and true hyphae are regularly observed during infection and have distinct functions (as discussed below), the role of pseudohyphae and switching in vivo is rather unclear and chlamydospores have not been observed in patient samples. Phenotypic plasticity (switching) has been proposed to influence antigenicity and biofilm formation of *C. albicans*. Yeast cells adhere to host cell surfaces by the expression of adhesins. Contact to host cells triggers the yeast-to-hypha transition. The expression of invasins mediates uptake of the fungus by the host cell through induced endocytosis. The fungus expresses specialized proteins on the cell surface (invasins) that mediate binding to host ligands thereby triggering engulfment of the fungal cell into the host cell. Als3 and Ssa1 bind to host E-cadherin and likely induce endocytosis by a clathrin-dependent mechanism.

Adhesion by host recognition biomolecules and secretion of fungal hydrolases has been proposed to facilitate the second mechanism of invasion, i.e., fungal-driven active penetration into host cells by breaking down barriers. Agglutinin-like sequence (ALS) proteins are arguably the best studied *C. albicans* adhesins which form a family consisting of eight members (Als1−7 and Als9). The ALS genes encode glycosylphosphatidylinositol (GPI)-linked cell surface glycoproteins. The hypha-associated adhesin Als3 is the most crucial for adhesion [6].

Following adhesion to host cell surfaces and hyphal growth, *C. albicans* hyphae can secrete hydrolases, which have been proposed to facilitate active penetration into these cells. Three different classes of secreted hydrolases are expressed by *C. albicans*: secreted aspartic protease (i.e., Sap1−10), phospholipases (PLPA-D), and lipases (LIP1−10), respectively.

A further important virulence factor of *C. albicans* is its capacity to form biofilms on abiotic or biotic surfaces. The attachment of yeast cells to abiotic (e.g., catheters) or biotic (host cells) surfaces can give rise to the formation of biofilms with yeast cells in the lower part and hyphal cells in the upper part of the biofilm. Several transcription factors control biofilm formation. These include the transcription factors Bcr1, Tec1, and Efg1. Deletion of any of these regulators (BCR1, TEC1, EFG1, NDT80, ROB1, or BRG1) resulted in defective biofilm formation in in vivo rat infection models [7].

An important environmental cue that triggers hypha and biofilm formation in *C. albicans* is contact sensing. Upon contact with a surface, yeast cells switch to hyphal growth. On surfaces with particular topologies (such as the presence of ridges) directional hyphal growth (thigmotropism) may occur [8].

In addition to these virulence factors, several fitness traits influence fungal pathogenicity. They include a robust stress response mediated by heat shock proteins (Hsps); auto-hyphal formation through uptake of amino acids, excretion of ammonia (NH_3), and concomitant induction of extracellular alkalization; metabolic flexibility and uptake of different compounds as carbon (C) and nitrogen (N) sources; and uptake of essential trace metals, e.g., iron (Fe), zinc (Zn), copper (Cu), and manganese (Mn) [9].

GENOME AND PROTEOME

C. albicans is infecting the ever-increasing immunocompromised patients globally and so has been of great interest to the scientific community. The genetics of this fungal pathogen is quite multifaceted paralleled to the baker's yeast *Saccharomyces cerevisiae*, and "classical genetics" has suffered a great setback in studying this organism. However, the accessibility of complete genome sequence has unlocked massive extent of opportunity for *Candida* community to study it with "reverse genetics" approach using advanced molecular genetics technology, proteomics, and genomics tools.

Moreover, *C. albicans* possesses most of the characteristics of *S. cerevisiae* with a shared similarity of 80% of genes in both the organisms. The genome of *C. albicans* is quite dynamic and lot of truncations, translocations, and other mutational events occur more frequently compared to other microbes. However, the genome of sequencing strain SC5314 published in 2004 was found to be quite stable having eight distinct chromosomes in duplicate ranging from 1030 to 3200 kb [10]. The genome size of *C. albicans* is estimated to be 14.3 Mb, and it contains about 6107 protein-coding genes. Of the 6107 genes/open reading frames (ORFs), about 774 are specific to *C. albicans* and homologues for these genes/ORFs are not available in *S. cerevisiae*. Effort has been made to sequence the genomes of other less-infectious *Candida* strains, and the complete genome sequences for *C. albicans* (WO-1), *C. dubliniensis*, *C. tropicalis*, *C. parapsilosis*, *C. guilliermondii*, and *C. lusitaniae* are available now. Gene annotation data available in CGD shows that the functions of only 22.97% (1403 genes) of the genes have been experimentally verified, whereas 77.03% (4705 genes) of the genes remain uncharacterized in *C. albicans* and their functions have been assigned on the basis of sequence analysis. Furthermore, 152 genes/ORFs are still in the "dubious" category for which no experimental evidence is available and seems to be indistinguishable from noncoding sequences [10]. The sequencing of other *Candida* species along with *C. albicans* has laid a platform to relate the genetic profile of these organisms and discover potential genes whose products are involved in adhesion, propagation, colonization, and survival. Furthermore proteomics is the gateway towards understanding vital functions of *Candida* in the area of infection and resistance. *Candida* proteome studies have been major contributing factors in revealing the virulence and pathogenicity factors, resistance towards drugs, and changes in cellular level in response to stress. Thus it plays significant role in understanding the utter complex genome [11].

GENETICS

C. albicans is a diploid organism with no known haploid phase unlike most yeast, and it was considered to be asexual for a long time. But genome sequencing has strongly transformed our understanding of this organism. *C. albicans* genome sequencing data has further revealed a mating-type (MAT-like) locus that led to the engineering of mating-competent strains. Advances in this area led to identification of a natural mating-competent form which mates naturally at high frequency to give a tetraploid gamete. So far, efforts have been unsuccessful to reveal meiosis, and thereby a complete sexual cycle. Although genes present in yeast *S. cerevisiae* that supposedly carry out the essential stages of meiosis have been found to be homologous to a huge repertoire of genes in *C. albicans* genome [12].

Because it is diploid and lacks a complete sexual cycle, conventional genetic analysis is simply not possible. Moreover, for more than a century, *C. albicans* was thought to be "imperfect" or asexual, based largely on the inability to identify mating forms in the laboratory. A striking characteristic of the *Candida* genus is that in many of its species like *C. albicans* and *C. tropicalis*, the CUG codon specifies serine which in the normal case codes for leucine in general. This is an uncommon example of diverging from the standard genetic code. Under certain environment this modification helps these *Candida* species by inculcating a permanent stress response.

METABOLOME

Metabolic studies have essential roles in understanding metabolic activity of yeast cells under normal as well as in stress conditions like starvation, alkaline pH, oxygen depletion, and high temperatures. Depending upon environmental conditions *C. albicans* have the extraordinary capacity to develop into several distinctive morphological forms: yeast, hyphae, and pseudohyphae. This ability is considered to play a key role as critical virulence factor. We aim to unravel the use of metabolomics and metabolic flux analysis in antifungal drug discovery. Moreover, metabolic pathways essential for *C. albicans* morphogenesis have been extensively studied. Morphogenetic process in this pathogen is mainly controlled by Ras, PKA, MAPK, HOG, and RIM101 signaling pathways, and regulated by Cph1p, Czf1p, Efg1p, Gcn4p, Mig1p, Nrg1p, Rfg1p, Ssn6p, Tec1p, Tup1p, and Ume6p transcription factors [13].

CHAPTER 2

Host—Pathogen Interaction

MORPHOGENESIS

Candida species are capable of growth as yeast, pseudohyphal or hyphal forms. When *C. albicans* infect humans and animals, hyphae predominate at the primary site of infection in epithelial layers and tissue, whereas yeast forms are found on the epithelial cell surface or merging from penetrating hyphae in surrounding tissue [14]. The cell wall of *C. albicans* contains cell wall proteins (CWPs) and carbohydrates are absent from the host. The cell wall therefore symbolizes a model immunological target to differentiate self from nonself and therefore the majority of fungal pathogen-associated molecular patterns (PAMPs) that trigger and regulate immune responses are mainly cell wall components. In well-preserved images of the *C. albicans* cell wall, two main layers are identified: an outer layer of glycoproteins and an inner layer of skeletal polysaccharides. The outer layer of *C. albicans* cell wall predominantly consists of *O*- and *N*-linked mannose polymers (mannans) covalently associated with proteins to form glycoproteins. Cell wall in totality contains 80—90% of only carbohydrates [15]. The chitin content of the cell accounts for 2—4% of the weight of the cell wall, and during the switch three- to fourfold increase in the chitin level is reported. The expression of CWPs is highly modulated during the yeast-to-hypha transition, and *Hwp1*, *Hyr1*, and *Als3* are highly upregulated genes encoding hypha-specific proteins. Cytoplasmic immunodominant antigens that are not normally associated with the cell wall have also been recognized. Nevertheless, the hypha-specific CWPs are major antigens and in addition they function as adhesins and invasins that can modulate immune responses [16]. Human pattern recognition receptors (PRRs) recognize not only fungal PAMPs but also few damaged host cell components, damage-associated molecular patterns (DAMPs). These DAMPs recognize components like nucleic acids and alarmins [17].

Anticandidal Agents. DOI: http://dx.doi.org/10.1016/B978-0-12-811311-0.00002-8

IMMUNE RESPONSES

Mammalian immune system triggers innate recognition of fungi, *C. albicans*, and it was able to elicit two different responses by dendritic cells (DCs) when phagocytosed in yeast or hyphae form. Whenever yeast cells are phagocytosed, DCs began a typical antifungal immune response, but hyphae cells are able to break out of the phagosome of DCs. Recognition of unopsonized *C. albicans* by macrophages and monocytes is arbitrated by recognition systems that sense *N*- and *O*-linked mannans, glucans, and mannosides [18].

Delayed-Type Hypersensitivity

Adaptive immune responses to *C. albicans* occur in immunocompetent individuals by delayed-type hypersensitivity. It is also called as Type 4 Hypersensitivity. Unlike the other types, it is not antibody mediated but rather is a type of cell-mediated response, i.e., it includes T cells, monocytes, macrophages, and cytokines in response to an antigen. This takes 12−48 h to fully develop after contact with the antigen. This response involves the interaction of antigens with the surface of lymphocytes. Sensitized lymphocytes can produce cytokines, which are biologically active substances that affect the functions of other cells [19].

During a fungal infection, the immune response must eliminate the fungus to restore a homeostatic environment while reducing the collateral damage to tissues. Mechanisms that mammalian hosts have developed to coexist peacefully with their microbiota can be manipulated by fungi to infect. Manipulation of the regulatory network of the host by the *Candida* is one such mechanism to ensure fungal survival [20].

Pattern Recognition Receptors

The first encounter occurs at the epithelium where *Candida* colonizes and gradually invades. As yeast transitions to the invasive hyphal morphotype, they exhibit a greater capacity for endocytosis and damage of epithelium, which causes the release of immunogenic cell wall constituents and pro-inflammatory cytokines and chemokines from the epithelium. Cytokine and chemokine release acts as the initial trigger for attracting monocytes and neutrophils in the circulation, as well as tissue macrophages. These cells express a repertoire of PRRs [14]. Depending on the fungal species and on the host cell types different PRRs are stimulated by fungal PAMPs (Fig. 2.1).

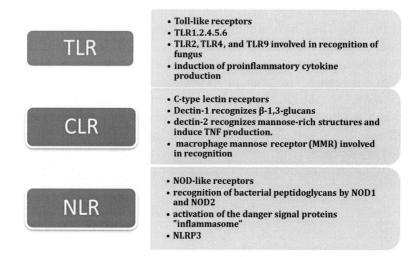

TLR
- Toll-like receptors
- TLR1.2.4.5.6
- TLR2, TLR4, and TLR9 involved in recognition of fungus
- induction of proinflammatory cytokine production

CLR
- C-type lectin receptors
- Dectin-1 recognizes β-1,3-glucans
- dectin-2 recognizes mannose-rich structures and induce TNF production.
- macrophage mannose receptor (MMR) involved in recognition

NLR
- NOD-like receptors
- recognition of bacterial peptidoglycans by NOD1 and NOD2
- activation of the danger signal proteins "inflammasome"
- NLRP3

Figure 2.1 A list of classes and functions of host cells expressing (PRRs) including TLRs, CLRs, and NLRs that recognize C. albicans *PAMPs.*

TH Cell Response

Adaptive immunity to fungi is only partially understood. The activation by PAMPs also initiates the subsequent pattern of T-cell activation. The effector T helper (TH) cells develop from proliferating TH cells and subdivide further into type 1 and type 2 cells also known as TH1 and TH2. TH1 cell activation is determined by the response of DC with the combination of TLR and CLR signals provided by *Candida*. TH1 cells are involved as a help in production of opsonizing antibodies. They are triggered by interleukin (IL)-12, IL-2 and their effector cytokine is interferon (IFN)-γ and tumor necrosis factor-β. IFN-γ secreted by CD4 T cells can activate macrophages and iNOS (inducible nitric oxide synthase) to produce free radicals to directly kill the pathogen. Inflammatory DCs initiate antifungal TH1 and TH2 cell responses that include TLRs. Several clinical observations suggest an inverse relationship between IFN-γ and IL-10 production in patients with fungal infections. High levels of IL-10, which negatively affect IFN-γ production, are detected in chronic candidal diseases. TH2 cells are positively triggered by IL-4 and their effector cytokines are IL-4, IL-5, IL-9, IL-10, and IL-13. TH II cells appeared to be producing cytokines that were turning off the fungicidal effector capabilities.

Th17 cells are a subset of TH cells developmentally distinct from Th1 and Th2 lineages producing IL-17A, IL-17F, IL-21, IL-22, granulocyte-macrophage colony-stimulating factor, and many other factors. They have an important function in the host response against extracellular pathogens, but they are also associated with the pathogenesis of many autoimmune and allergic disorders. TH17 cell activation occurs in fungal infections, mainly through the SYK−CARD9, MYD88, and mannose receptor signaling pathways in DCs and macrophages. TH17 cells are present in the fungus-specific T cell memory repertoire in humans and involved in both TH1- and TH2-type responses [16].

CHAPTER 3

Antifungals Used Against Candidiasis

An antifungal agent kills fungi or inhibits their growth. Antifungals that kill fungi are called fungicidal while those that merely inhibit their growth are called fungistatic. Fungal infections mostly infect the mucus membranes. Superficial infections like yeast vaginitis, oral thrush, affect the skin or mucous membranes affect millions of people worldwide. But these are rarely life-threatening. Most superficial fungal infections are easily diagnosed and can be treated effectively. Invasive infections occur when fungi get into the bloodstream and generally cause fatal effects. These are also called systemic infections. They may be caused either by opportunistic fungal pathogens like *Candida* that attacks on weakened immune system or by an invasive organism like *Cryptococcus* that is common in a specific geographic area. Unlike superficial infections systemic fungal infections can be life-threatening like pulmonary aspergillosis, blastomycosis. Systemic antifungals are a choice for patients with acute cases of *Candida* infections and in immunocompromised patients. Topical antifungals are used for simple, localized candidiasis in patients with normal immune function. These antifungals could be synthetic or could be derived from natural sources. The first landmarks in the development of active and safe antifungal agents were the discovery of the antifungal activities of griseofulvin by Oxford, in 1939. Infections caused by pathogenic *C. albicans* are commonly treated with mainly azoles and other nonazole antifungal agents, but they have evident side effects. The first azole was given by Wooley in 1944. Excessive use of synthetic antifungals causes liver damage, allergic reactions altered estrogen levels, and development of drug resistance.

Antifungal medicines can be grouped according to their source, structure, and mode of action. Based on those, antifungal drugs can be divided into five major classes (Fig. 3.1).

Anticandidal Agents. DOI: http://dx.doi.org/10.1016/B978-0-12-811311-0.00003-X

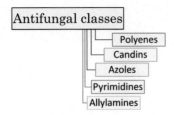

Figure 3.1 Classification of anticandidal agents.

AZOLES

They represent the most common class of antifungal used. Azoles that are available for systemic use can be categorized into two categories— Imidazoles and Triazoles. Triazole antifungals include Fluconazole, Itraconazole, Posaconazole, Voriconazole whereas Imidazole antifungals include Clotrimazole, Econazole, Miconazole, Ketoconazole, and Tioconazole. Few literatures also report another class called Thiazoles that includes Abafungin.

Activity and Mechanism of Action

The mode of action of azole antimycotics is by inhibition of cytochrome P450-dependent lanosterol 14α-demethylase, an enzyme involved in ergosterol biosynthetic pathway and this suppression results from binding of nitrogen of the heterocyclic portion of the azole molecule (like imidazole, 1,2,4-triazole, pyrimidine or pyridine) to the heme iron of cytochrome P-450. In addition to accumulation of methylated sterols, there is also a decrease in unsaturated to saturated fatty acid ratio and there is also a shift from C_{18} to C_{16} fatty acids in vivo. Any impairment in the synthesis of membrane sterol affects normal growth and drug susceptibility. Recent findings have also suggested the involvement of ergosterol in the formation of hyphae, one the virulent attributes of *C. albicans*. Ergosterols maintain the membrane rigidity, stability, membrane integrity and also provide resistance to physical stresses and loss of sterol leads to destabilization of the membrane leading to increased permeability and altered drug susceptibility (Fig. 3.2).

Azole Variants and Their Structure

Imidazoles are predominantly used topically. They have a two-nitrogen azole ring in their structure. They have been replaced for

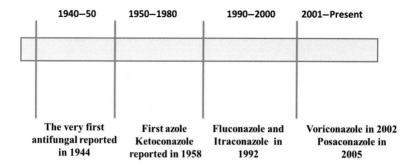

1940–50	1950–1980	1990–2000	2001–Present
The very first antifungal reported in 1944	First azole Ketoconazole reported in 1958	Fluconazole and Itraconazole in 1992	Voriconazole in 2002 Posaconazole in 2005

Figure 3.2 Development of azoles over the time.

systemic administration by Triazoles, which have three nitrogen in the azole ring. The commonly known imidazole compounds are clotrimazole, ketoconazole, and miconazole. Azole derivatives (e.g., fluconazole, itraconazle, voriconazole, ravuconazole, posoconazole) with 1, 2, 4-triazole ring in place of imidazole have better pharmacokinetics profile with better fungicidal activity and lower toxicity.

Ketoconazole
This imidazole compound was first available as broad-spectrum antifungal. It was effective against systemic and dermatophytic infections and most fungal infections. Initially Ketoconazole transformed the treatment against infection but later it was replaced due to adverse gastrointestinal side effects.

Ketoconazole

Itraconazole
It is a triazole compound. This antifungal was the first azole in humans, not only effects against oral candidiasis but also exhibits significant activity against *Aspergillus*.

Itraconazole

Fluconazole

Fluconazole (FLU) is major azole drug used as therapeutic as well as prophylactic treatments. It has relatively few drug interactions and is safe but lacks activity against filamentous fungi. Hence major resistance is found against pathogenic *Candida* due its filamentation property.

Fluconazole

Voriconazole

It is the second-generation azole used as first-line Itreatment of invasive esophageal candidiasis. These are functional against many species of *Candida* including strains resistant to FLU and also effective against *Aspergillus*.

Voriconazole

Posoconazole

It is second-generation antifungal triazole and structurally similar to that of itraconazole. It is the last addition to this group and has better activity than voriconazole against *Aspergillus, Candida, Cryptococcus,* and *Histoplasma.*

Posaconazole

Ravuconazole

Ravuconazole is highly active against broad spectrum of fungi including most species of *Candida* and few isolates that were resistant to FLU. It is still in final stages of clinical trials.

Ravuconazole

Terconazole

Terconazole is a triazole ketal with broad-spectrum antifungal/antimycotics tendencies.

Other Notable Triazoles Include

The earlier imidazole derivatives (such as miconazole, econazole, and ketoconazole) have a complex mode of action, inhibiting several membrane-bound enzymes as well as membrane lipid biosynthesis. An accumulation of zymosterol and squalene synthesis was observed when *C. albicans* cells were treated with voriconazole. It is unclear whether the accumulation of these intermediates results from voriconazole interaction with various (non-14α-demethylase) enzymes involved in ergosterol synthesis or from secondary effects of 14α-demethylase inhibition [21]. Triazoles have broader spectrum than imidazoles. Except ketonazole the other members are restricted to treatment of superficial infections [22].

POLYENES

Activity and Mechanism of Action

The mechanism of action of the polyenes has been associated with their ability to bind to the membrane sterols resulting in the formation of pores that disrupt the stability and structure integrity of cell membrane.

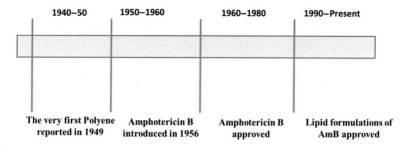

Figure 3.3 Development of polyenes over the time.

This is manifested by increase in cell permeability, leakage of cyto-plasmic contents such as potassium, magnesium ions, sugars, metabo-lites resulting in cell death. Although the affinity of polyenes for fungal ergosterol is much higher, it can also interact with mammalian sterols (cholesterol). Nephrotoxicity is the major adverse effect limiting the use of amphotericin B. Generally, amphotericin B has a very broad range of activity and is active against most pathogenic fungi. On occasion, how-ever, isolate of any species may be found to be resistant. Amphotericin B and its lipid formulations remain as the mainstay of antifungal ther-apy. Its lipid formulations, on the other hand, are promising due to their ability to reduce the toxicity of amphotericin B. The use of lipid formu-lations in specific clinical settings is under continuing investigation. Liposomal nystatin is in late Phase III clinical trials (Fig. 3.3).

Structure and Variants
Structurally they have a macrolide ring, composed of 26–38 carbon atoms and unsaturated carbon atoms and hydroxyl groups. These fea-tures of the molecule contribute to the polyenes' amphipathic proper-ties (those relating to molecules containing groups with different properties, e.g., hydrophilic and hydrophobic). The number of conju-gated double bonds varies with each polyene, and the compounds are generally classified according to the degree of unsaturation [23].

Amphotericin B
Amphotericin B (AmB) *"The gold standard of polyenes"* is the most effective antifungal and is still effective for over three decades. They directly bind to the membrane ergosterol resulting in pore formation which makes the cell membrane leaky and results in cell death. It was

isolated from bacteria *Streptomyces noursei* in 1956. It is fungicidal in nature, one of the best drugs developed against *Candida* but limited to fungus with sterols only. AmB causes oxidative damage to plasma membranes. In higher concentrations, polyenes also inhibit chitin synthase, a cell wall synthetic enzyme localized in the membrane. The interaction between AmB and human cell membranes containing cholesterol results in toxic side effects of the drug [24].

Nystatin

It is very much similar to Amphotericin B in terms of structure and mode of action. It was discovered in 1950 and is still used as a topical antifungal agent. It is effective as topical treatment of oropharyngeal candidiasis but the problem with its usage is that it has been reported to be nonabsorbable after oral administration.

Natamycin

They directly bind to the membrane ergosterol resulting in pore formation which makes the cell membrane leaky and results in cell death [25].

ECHINOCANDINS

Activity and Mechanism of Action

Candins or Echinocandins are compounds that inhibit cell wall synthesis by inhibiting the synthesis of the structural polymer β-1, 3-glycan. Beta-glucan destruction prevents resistance against osmotic forces, which leads to cell lysis. Treatment of fungi with β-glucan synthase inhibitors causes noncompetitive inhibition of 1, 3, β-glucan synthase with secondary effects on other constituents, such as an increase in the chitin content of the cell wall and a reduction in the ergosterol content of the fungal cell membrane [26]. They are sometimes referred to as "Penicillin of antifungals." Caspofungin, micafungin, and anidulafungin are derivatives of echinocandins of semisynthetic nature with emerging use due to their solubility, antifungal spectrum, and pharmacokinetic properties. The fact that resistance developed against this class of antifungals is rare makes it preferable. Also they are effective against FLU-resistant strains and fungicidal against most species of *Candida*.

Echinocandins Variants

Caspofungin

It is a lipopeptide antifungal drug that is administered intravenously. It has potent activity on invasive infections. It was found to have notable effects against cell wall. Caspofungin has few significant interactions as it is neither a substrate nor an inhibitor of the cytochrome P-450 system. It blocks an enzyme that catalyzes the polymerization of uridine diphosphate-glucose into β (1−3) glucan.

Micafungin

It was approved for the treatment of oral candidiasis in the 2000s. It is also administered intravenously for prophylaxis treatment in patients of candidemia, acute disseminated candidiasis, *Candida* peritonitis, abscesses, and esophageal candidiasis.

Anidulafungin

Anidulafungin is a semisynthetic candin. It has proven efficacy against esophageal candidiasis, but its main use will probably be in invasive *Candida* infection.

Pneumocandins

Pneumocandins are lipohexapeptides of the echinocandin family that potently interrupt fungal cell wall biogenesis by noncompetitive inhibition of 1,3-β-glucan synthase [27].

Pneumocandin B$_0$

Pneumocandin C$_0$

Echinocandin B

It was the first of the echinocandin class of antifungals. A lipopeptide is a naturally occurring cyclic hexapeptide with a linoleoyl side chain. Echinocandin B is a fermentation product of *Aspergillus nidulans* and *Aspergillus rugulosus* [28].

Cilofungin

It is the first clinically applied member of the Echinocandin family of antifungal drug. It was derived from a fungus in the genus *Aspergillus* [29].

ALLYLAMINES

Activity and Mechanism of Action

These antifungal agents are reversible, noncompetitive inhibitors of blocking the ergosterol biosynthetic pathway and leading to accumulation of toxic squalene. First step of the biosynthesis is the conversion of squalene to squalene-2, 3-epoxide by squalene epoxidase. The increase of squalene in the cell membrane is toxic to the cell, causing pH imbalances and breakdown of membrane-bound proteins [30].

Variants of Allylamines

Terbinafine

It was used only for topical treatment of superficial mycoses, and neither of its targets has attracted recent research interest. It has been commercially available since May 1996. Terbinafine has good in vitro activity against filamentous fungi but has variable activity against yeasts. Terbinafine has been shown to be fungicidal against dermatophytes and dimorphic fungi.

Terbinafine

Naftifine

It is an allylamine antifungal drug for the topical treatment of fungal infections. Its precise mechanism of action involves selectively blocking sterol biosynthesis via inhibition of the squalene 2, 3-epoxidase enzyme.

PYRIMIDINES

Activity and Mechanism of Action

They inhibit nucleic acid synthesis by inhibiting thymidylate synthase. Thymidylate synthase methylates deoxyuracilmonophoshate (dUMP) into deoxythyminemonophosphate (dTMP). Fluoropyrimidines, such as fluorocytosine (5-FC) and 5-fluorouracil (5-FU), are synthetic structural analogs of the DNA nucleotide cytosine. 5-FC itself has no antifungal activity, and its fungistatic properties are dependent upon the conversion into 5-FU. The drug rapidly enters the fungal cell through specific transporters, such as cytosine permeases or pyrimidine transporters, before it is converted into 5-FU by the cytosine deaminase [11]. These are also included in the category of *Antimetabolites*.

Pyrimidine Variants
5-FC or Flucytosine
It is synthetic water-soluble pyrimidine analog that inhibits DNA synthesis. They are often used in synergism with Amphotericin B or Itraconazole. Amphotericin B increases cell permeability allowing more FC to penetrate the cell. It is fungistatic in nature effecting not only *Candida* but also *Cryptococcus*.

5-Fluorocytosine

5-Fluorouracil

THIOCARBAMATES AND MORPHOLINES

Thiocarbamates (e.g., Tolnaftate) inhibit early step in ergosterol biosynthesis by blocking enzyme squalene epoxidase encoded by *ERG1*.

Tolnaftate
Although the exact mechanism of action is not entirely known, it is believed to inhibit squalene epoxidase and structurally similar in the class of allylamines.

Morpholines

They inhibit Erg2 that catalyze catalyzing sterol D14-reductase and D8−D7 isomerase (Erg24). Although thiocarbamates, morpholines, and allylamines have broad spectrum against fungal species, they are usually deployed as topical agents to treat dermatophyte infections [31].

OTHER INHIBITORS

Cell Wall Synthesis Inhibitors

Mannoproteins are assembled, transferred to asparagine residues in proteins, and modified by glycosidases in a similar manner in yeast and mammals. During intracellular maturation, mannoproteins are coupled to GPI group but this membrane anchor is replaced by β-(1, 3)-glucan in yeast cells. This step being unique in yeasts is identified as potentially selective targets for antifungals. The two classes, benanomicins and paramicin, appear to target cell wall mannoproteins [32].

Chitin Biosynthesis Inhibitors

Chitin being an important component of fungal cell wall has been a potential drug target. Inhibition of chitin synthase enzymes by structural analogs of the substrate of chitin biosynthesis, UDP-*N*-acetylglucosamine turned out to be fungistatic and in some cases fungicidal. Polyoxins and nikkomycins are two classes of nucleoside-peptide antibiotics which are competitive inhibitors of chitin synthase. Nikkomycins have also displayed synergistic activity with some azoles and glucan synthase inhibitors. But *C. albicans* have shown resistance to polyoxins mainly due to their poor transport through cell membrane. Modification of these to produce drug of clinical values have failed till now.

Ergosterol Biosynthesis Inhibitor

These lead to reduced ergosterol biosynthesis and are thus conceptually related to the azole antifungal agents. One example is terbinafin, which acts by inhibiting squalene epoxidase, the enzyme mediating the conversion of squalene to 2, 3-oxido-squalene, and the first step of ergosterol biosynthesis. Among other ergosterol biosynthesis inhibitors, some morpholine analogs like fenpropimorph and amorolfine have inhibition of sterol biosynthesis as primary site of action [33].

Glucan Synthesis Inhibitors

Agents of this class inhibit glucan synthesis. Glucan is a key component of the fungal cell wall, and inhibition of this enzyme produces significant antifungal effects. Two classes of these inhibitors are papulacandins and echinocandins. Chemical modifications of echinocandins including changes in the peptide core have resulted in analogs with improved efficiency. Example: Caspofugin, micafugin, anidulafugin of which only caspofugin is approved for clinical use, rest are on clinical trial stage [34] (Table 3.1).

Table 3.1 List of All the Drugs, Their Mode of Action, and Resistance Mechanism			
Drugs	**Mode of Action**	**Resistance**	**Administration**
Polyenes			

1. Amphotericin B
  1.1. Amphotericin B deoxycholate
  1.2. Amphotericin B lipid complex
  1.3. Amphotericin B colloidal dispersion
  1.4. AmphotericinB liposomal
2. Nystatin
	Destabilizes the fungal cell membrane by increased permeability of cell membrane ergosterol and oxidative damage	Induction of low membrane ergosterol content detected in some fungi	Mostly intravenously but oral solution also

Azoles			

1. Triazoles
  1.1. Fluconazole
  1.2. Itraconazole
  1.3. Voriconazole
  1.4. Posoconazole
2. Imidazoles
	Inhibition of Erg11p-CYP-dependent C-14α demethylase that converts lanosterol to ergosterol; conversion of Erg11p substrate into toxic methylated sterols	Overexpression of Erg11p, efflux via ATP-binding cassette (ABC) and major facilitator superfamily (MFS) transporters, mutation in *ERG3*, import of host cholesterol stress tolerance induction	Oral tablet
			Oral capsule
			And
			Oral suspensions

Allylamines			
1. Terbinafine 2. Naftifine	Inhibition of squalene epoxidase; accumulation of toxic sterol intermediates	Mutations in Erg1p, efflux via ABC pumps, stress tolerance induction	
Pyrimidine analogs			
1. 5-FC or Flucytosine	Interferes with RNA and DNA synthesis by misincorporation of 5-FU or cytosine	Mutation in Fur1p, i.e., uracil phosphoribosyl transferase	Oral tablet
Echinocandins			
1. Caspofungin 2. Micafugin 3. Anidulafungin	Blocking cell wall biosynthesis by inhibition of $\beta(1,3)$-glucan synthase thus lysis of cell	Mutation in $\beta(1,3)$-glucan synthase, loss of heterozygosity for $\beta(1,3)$-glucan synthase	Intravenously

NOVEL INHIBITORS

Expected Drugs (Compounds in Pipeline)

Antifungal drug spectrum offers rather limited choices and the efficacy of therapy particularly in immunosuppressed persons is strongly affected by the host factors as well. A long time is invested for the differential and careful development of these drugs. Some investigational compounds in Phase 1−2 / preclinical trials are given below.

SCY078

Formerly known as MK-3118 is an Enfumafungin derivative, a new glucan synthase inhibitor currently in Phase 2 trial. It exhibited increased activity against Echinocandin-resistant *Candida* species in vitro [35,36].

Ara-sertaconazole Nitrate

Different formulations of Ara-sertaconazole nitrate are currently in Phase 2 trial of development. It exhibited elevated in vitro activity against FLU-resistant *Candida* species. It is the R-enantiomer of a successful antifungal developed sertaconazole nitrate which has been marketed for many years. It is used for treatment of vulvovaginal candidiasis. The racemate itself is also used as a broad-spectrum antifungal drug.

VT-1161

This one is an oral inhibitor which is highly selective for inhibition of fungal molecule CYP51 (Sterol 14 alpha-demethylase). In vitro and in vivo studies demonstrate broad-spectrum activity against both *Candida* species and dermatophytes, including those species that cause recurrent vulvovaginal candidiasis (RVVC) and onychomycosis [37].

MGCD290
This compound is a Hos2 histone deacetylase inhibitor. It has effects in combination with Echinocandins against Echinocandin-resistant *Candida* species in vitro. It is also given in combination with azole antifungals against opportunistic pathogens. In vitro, MGCD290 in combination with fluconazole reverses fluconazole resistance (primary and acquired) in a wide range of fungal species, including *C. glabrata*. MGCD290 has completed multiple Phase I studies in healthy adult volunteers and has shown an excellent safety profile [38].

E1210
It is a compound in its preclinical trial. It is basically an Inositol acyltransferase inhibitor of fungi. It is active against *Candida* including FLU-resistant *C. neoformans* and *Aspergillus* species [39].

T-2307
It is a novel arylamidine and has in vitro and in vivo broad-spectrum activities against pathogenic fungi and its uptake might be mediated by a transport system. T-2307 exhibits excellent antifungal activity against fluconazole-resistant *Candida* spp. Since T-2307 has a novel arylamidine structure and an antifungal spectrum different from those of azoles, candins, and polyenes, we hypothesized that T-2307 possesses a novel mechanism of action [40].

A

ASP9726
It is a novel second-generation echinocandin with improved activity. This compound is present in preclinical trial of its development. It has potent *Aspergillus* hyphal growth inhibition and significantly improved MIC against *C. parapsilosis* and echinocandin resistant-*Candida* [41].

Biafungin (CD101)
It is an Echinocandin administered intravenously. It has potent activity against *C. albicans*, *C. glabrata*, and *C. tropicalis* [42].

Isavuconazole
It is a triazole antifungal drug. Its prodrug, isavuconazonium sulfate, was granted approval by the U.S. Food and Drug Administration just recently in 2015. It was approved for treatment of invasive aspergillosis and invasive mucomycosis but its potential effects in *Candida* are still under testing.

Albaconazole

It is a triazole antifungal. It has potential broad-spectrum activity. It demonstrated considerable efficacy in animal models like *Aspergillus, Candida, Cryptococcus, Scedosporium* but no further development.

BQM—Bis [1, 6]quinolizinium 8-Methyl-Salt

C. albicans overexpressing (multidrug resistance-1) MDR1 exhibits highly increased susceptibility to a novel small-molecule compound, BQM, having the most potent antifungal activity. BQM could exploit the drug transporter Mdr1p to increase the accumulation, but it was uncertain if efflux pumps are inhibited by BQM or not [43].

Quinazolines

Quinazolines and its derivative compounds are broad spectrum in nature and have good antifungal activity against *C. albicans*. Derivatives of 4(3H)-quinazolinone were synthesized and evaluated for antifungal activity against *C. albicans*. The results were shown to have potent activity against pathogenic fungi.

Indole and its Derivatives

1H-Indole-4, 7-diones were synthesized and tested for in vitro antifungal activity against *Candida*. It appears that a 2-thioxothiazolidin-4-one ring and an enone linkage are indispensable for activity [44].

Few molecules are still in clinical development like *Nanoparticle formulations of AMB and itraconazole*. They are in earlier phases of drug development.

Inhibitor Molecules

Sordarins

Protein synthesis inhibitors include sordarins which selectively inhibit fungal protein synthesis by blocking the function of elongation factor 2 (EF-2) and ribosomes. They are absent in human cells.

Sphingolipid biosynthesis inhibitors are also being currently investigated to identify new antifungal targets. Recent studies show that DNA topoisomerases are apparently suitable targets for drugs, e.g., eupolauridine, a potential inhibitor of topoisomerase has an advantage of being nontoxic to mammalian cells.

Tacrolimus

It was formerly known as FK506; the 23-membered natural product (NP) macrolide lactone is involved in blocking T-cell activation. Its mode of action is similar to cyclosporins but they are structurally unrelated. It blocks calcineurin, a Ca^{2+}-calmodulin-dependent serine-threonine protein phosphatase subsequently blocks Calcium-dependent events, such as IL-2 gene transcription, nitric oxide synthase activation, cell degranulation, and apoptosis. Calcium signaling is responsible for this pathogen in responding to several stresses.

Ascomycin

Also called immunomycin/FR-900520/FK520, it is an analog of tarcolimus isolated from fermentation broths of *Streptomyces* species. It inhibits the production of Th1 (INF- and IL-2) and Th2 (IL-4 and

IL-10) cytokines. Additionally, ascomycin preferentially inhibits the activation of mast cells.

Milbemycin

Milbemycin and its oxim derivatives are known as ABC transporter inhibitors are products of fermentation by *Streptomyces* species. It opens glutamate-sensitive chloride channels in neurons and myocytes of invertebrates, leading to hyperpolarization of these cells and blocking of signal transfer. The commercially available milbemycin A3/A4 has significant activity against ABC transporters of *C. albicans* and *C. glabrata*.

Unnarmicin A and Unnarmicin C
They are novel cyclic peptides isolated from the fermentation broth of a marine bacterium, *Photobacterium* sp. They were able to sensitize cells overexpressing azole drug pumps CaCdr1p with the potential to be used as adjuvants for antifungal chemotherapy.

Verapamil

It is an L-type voltage-gated calcium channel blocker that also effects on oxidative stress response in this fungus. In *Candida* oxidative agent H_2O_2 is associated with a decrease of calcium fluctuation under the stress. Verampil causes enhanced oxidative stress that increases levels of reactive oxygen species (ROS) and enhanced dysfunction of the mitochondria under the oxidative stress [45].

Disulfiram

It is a bis-disulfide and is basically an alcohol antagonistic drug that produces an acute sensitivity to ethanol. In scientific studies it was reported that Disulfiram has significant antifungal action against *P*-glycoprotein. It is reported to inhibit ATP hydrolysis hence exhibiting potential role in combating MDR.

Enniatin B

It is a bioactive compound and structurally is a cyclodepsipeptide. It reverses the fluconazole resistance by inhibiting CDR1p and also affects Rhodamine 6G (R6G) efflux from CaCDR1p.

Trichostatin A

Many other molecules have shown inhibitory effects against *Candida* like treatment of *C. albicans* with a histone deacetylase inhibitor called Trichostatin A. It makes *Candida* more sensitive towards the drugs and induction of efflux upon fluconazole treatment was reduced.

Levofloxacin and Derivatives

They were able to inhibit the biofilms of *C. albicans* the yeast-to-hyphal transition and were also able to disrupt their mature biofilms. Many natural compounds and their extracts are also exploited for their antimicrobial properties. One of the advantages of these extracts or plant oils are they render minimum side effects to the already compromised patients. When used in combination to already existing drugs they prove to be a great rescue against ever-growing problem of resistance.

Substances and extracts isolated from different natural resources especially plants have always been a rich arsenal for controlling the fungal infections and spoilage. The methanol extract of turmeric demonstrated antifungal activity against *C. neoformans* and *C. albicans*. *Curcumin* is a mixture of curcuminoids extracted from *Curcuma longa*. It is a polyphenolic compound containing phytochemicals that has variety of broad-spectrum activities including antibacterial, antiviral, antifungal, and antimalarial activities. It targets *ERG3* of *C. albicans*. Its effect is not only against various strains of *C. albicans* but also against *C. glabrata, C. krusei, C. tropicalis, C. guilliermondii,* and *C. dubliniensis* [46]. Extensive research have been under way in recent years to isolate natural inhibitors of MDR exporters, since NPs have the potential to yield a large number of new anticandidal bases and formulations [47]. There are a significant number of NP drugs in development.

DIFFERENTIAL EXPRESSION OF GENES IN RESPONSE TO DRUGS

Scientists have used fixation protocols, microarray, and other transcriptome analysis techniques to determine the effect on expression of genes after treatment with antifungals specifically azoles. Among genes found to be differentially expressed were those involved in amino acid and carbohydrate metabolism; cell stress, cell wall maintenance; lipid, fatty acid, sterol metabolism; and small molecular transport. *CDR1*, *CDR2*, and *CaMDR1* which have been earlier mentioned to be associated with azole resistance were found to be upregulated in the resistant isolates of *Candida*. In addition to the major *ABC* transporter gene involved in drug efflux mediated resistance *CDR1*, the ergosterol biosynthesis gene *ERG2* was the found to be differentially regulated. Also upregulation of various genes like *RTA3* gene (exact function unknown but its homologous genes are involved in drug resistance) and *CRD2*, *GPX1*, and *IFD5* involved in the cell stress genes were found reported by researchers. Several other genes like *ALD5* encodes mitochondrial aldehyde dehydrogenase, which is essential for the modulation of proper functioning of etc. Downregulation of *ALD5* may contribute to the overexpression of *CDR1* and *CDR2*. Each of these genes holds presumed role in antifungal drug resistance but only a slight study has been put up towards their specific function in this process. The constitutive upregulation of *CDR/MDR* is usually because of gain-of-function mutations in the zinc cluster transcription factors Upc2, Mrr1, and Tac1. Scientists have identified a zinc cluster transcription factor, designated as *MRR1* (multidrug resistance regulator), that was upregulated with *MDR1* in drug-resistant isolates of clinical *C. albicans*. Inactivation of *MRR1* in two such drug-resistant isolates abolished both *MDR1* expression and *MDR*. During biofilm production β-1, 6-glucan biosynthesis is the reason for formation of matrix that confers resistance against AMB. Two genes involved in β-1, 6-glucan biosynthesis *SKN1* and *KRE1* were found to be differentially expressed after in vitro exposure to antifungal treatment. These genes including *CDR1* were responsible for resistance in *Candida* biofilms under the effect of antifungal exposure.

Hyphal development genes are also negatively regulated by transcriptional repressors Tup1, Nrg1, Sfl1, and Rfg1. *ADR1* and *ZRT2*, both of which are putative zinc finger transcription factors, where *ZRT2* is also transcriptionally induced during interactions with macrophages.

In azole-resistant isolates upregulation of certain genes were seen for instance *ABC* transporters from *C. dubliniensis* (*CdCDR1* and *CdCDR2*), *C. krusei* (*ABC1* and *2*), *C. tropicalis* (*CDR1* homolog) and from *C. neoformans* (*CnAFR1*, antifungal resistance 1) were the ones reported. Genes involved in the assimilation of fatty acids and other alternative carbon sources like *ACO1, ACS1, CIT1, FAA4, MLS1, POX4, SDH12* were found to be downregulated. Two genes involved in metal ion transport *FET34* and *FTR2* were also found to be downregulated by the action of azoles [48]. GPI-anchored proteins are critical for cell wall maintenance and integrity. The GPI synthesis gene *GPI1* encodes a product involved in biosynthesis of cell wall was found to be downregulated whereas the other gene *CWH8* involved in mannoproteins layer generation was upregulated. *LYS21* and *LYS7* encode homocitrate enzymes involved in the first and second step of the pathway for the biosynthesis of lysine. For adhesion genes *ALS1, ALS2, ALS4,* and *CTA1, ENA22* stress adaptation genes were differentially expressed. The other down-regulated genes included *ECE1, HYR1, RBT5,* those involved with morphogenesis *CDC19, HGT11, HXK2, HXT5, HXT61, HXT62,* involved in fermentation *BEL1, RPL18, RPS13, RPS21,* protein biosynthesis and *ALS10, HYR1, IHD1, PGA54, PGA59, PGA10, PHR1, RBT5, SUN41* are the genes associated with the cell surface.

Upregulation of histone-encoding genes *HHF1, HTA1,* and *HTB1* is reported. A cytoskeletal gene *TUB2* and genes that encode subunits of the ribonucleotide reductase complex *RNR1* and *RNR22* were also upregulated [49–51].

CHAPTER 4

Drug Resistance in *Candida*

The evolution of drug resistance depends on genetic variability, the cause of which is mutation. If a mutation deliberating resistance develops in the pathogen, its outcome is determined by as alteration, migration, selection, genetic drift, and recombination. Understanding the evolution of drug resistance is essential to prolonging the efficacy of existing drugs and to strategically developing novel drugs [52].

EMERGENCE OF DRUG RESISTANCE

Prolonged and widespread usage of antifungals in recent years has led to the rapid development of MDR, since the cells have evolved elaborate molecular mechanisms to protect themselves from the injuries caused by environmental exposure to toxic compounds of different structures and functions [53,54]. Therefore, MDR is a multifactorial phenomenon where a combination of mechanisms could contribute to drug resistance. The resistance can be broadly grouped into clinical, cellular, and molecular mechanisms. Primary resistance occurs in organisms never exposed in that host to the drug of interest. In contrast, secondary resistance, also defined as acquired resistance, arises only after exposure of the organism to the drug. Intrinsic resistance is defined as resistance of all or almost all isolates of one species to a certain drug—e.g., the resistance of *Candida krusei* to fluconazole. In addition, the term "clinical resistance" can be used to describe failure of therapy or relapse of infection with an organism not associated with in vitro resistance—e.g., changes in the host's immune system, such as neutropenia [55]. The factors contributing to clinical antifungal resistance are drug susceptibility of the strain, yeast-to-hyphal transition or switch phenotype, genomic stability of the strain, fungistatic nature of the drug, dosing and pharmokinetics of the drug, immune status and severity of infection, patient compliance with the drug regimen, and site of infection. The cellular mechanism for drug resistance may

Anticandidal Agents. DOI: http://dx.doi.org/10.1016/B978-0-12-811311-0.00004-1

be by change to a more resistant species/strain of *Candida*, genetic alterations rendering the strain more resistant by random mutations and transient gene expression by which a cell can alter its phenotype to become resistant in the presence of the drug [56]. The molecular mechanisms of antifungal resistance include alterations in the drug import by change in sterol components of the plasma membrane, alterations in intracellular drug processing (modification and degradation), genetic changes by point mutations, over expression, gene amplification, gene conversion or mitotic recombination in target enzyme of azole drugs, i.e., *ERG11* (cytochrome P450 dependent lanosterol 14α-demethylase, an enzyme involved in ergosterol biosynthetic pathway), alterations in other enzymes of the ergosterol biosynthetic pathway, and increased efflux of drug from the cells by over expression of drug efflux pump proteins [57].

One significant obstacles preventing successful antifungal therapy is the dramatic increase in drug resistance especially towards azoles. This treat to antifungal therapy by development of antifungal resistance is conferred to a high degree by over expression of drug efflux pump proteins. There are two major classes of drug efflux pumps in *C. albicans* namely, ABC superfamily and MFS of transporters. Members of ABC superfamily hydrolyze ATP to efflux out drugs whereas MFS proteins are proton motive force (PMF)-dependent antiporters which efflux out substrates in exchange of one or more H^+ ions. The major ABC transporters in *C. albicans* are CDR1 and CDR2 whereas CaMDR1 and FLU1 are MFS transporters. CDR1 and CDR2 confer resistance to a number of clinically relevant drugs like azole (fluconazole, ketoconazole, and itraconazole) and allylamine (terbinafine, amoralfine) antimycotics. CDR2 also provides a specific resistance for crystal violet. CaMDR1 conferred resistance to tubulin-binding drug benomyl and dehydrofolate reductase inhibitor, methotrexate, in addition to cycloheximide (inhibitor of protein biosynthesis), 4-nitroquinoline-*N*-oxide (a mutagen), sulfomethuron methyl, and benztriazoles. FLU1 confers resistance to fluconazole, itraconazole, ketoconazole, and mycophenolic acid. CDR1, CDR2, and CaMDR1 have found to be over expressed in *C. albicans* azole-resistant clinical isolates indicating their involvement in clinical resistance to fluconazole, the most commonly used antifungal. Resistance to azoles has also been found to be accompanied by the increased expression of the target enzyme, *ERG11*. Moreover, fluconazole-resistant strains have also been found to be cross-resistant to ketoconazole, itraconazole, sometimes also to amphotericin B. An emerging mechanism for resistance is modification of membrane lipid composition in the strains. There are reports that actions of antifungals are influenced by subtle modification of membrane lipids. Clinical and adapted strains of *C. albicans* have been shown to exhibit altered membrane phospholipid and sterol composition. The various classes of lipids like ergosterol, phospholipids, fatty acids, sphingolipids, etc. can be exploited for developing membrane lipid components as attractive drug targets. Moreover membrane lipids have also been shown to affect the functioning of the drug efflux pumps as well as morphogenesis of *C. albicans*. The interplay between the membrane lipids and azole-resistance mechanism can be further studied for exploitation in chemotherapy [58]. This reduced intracellular concentration of drugs ensures minimum amount of drug available for its target hence ensuring cell survival even in the presence of drugs.

The increasing incidence of systemic *Candida* infections coupled with lack of potent antifungals and emergence of drug resistance strains in immunocompromised patients receiving long-term chemotherapy have provided the impetus for the need of better understanding of the molecular genetics and mechanisms of drug resistance. This would be important for study of selective drug targets in the design of novel antifungals. Antifungal currently used in clinics includes polyenes, azoles, and 5-FC [59]. Azoles have fewer side effects than amphotericin B. But there are problems of toxicity (itraconazole, ketoconazole), resistance to azoles which actually pave the way for the need of newer antifungal drug targets like sphingolipid biosynthetic pathway, DNA topoisomerases, and protein translational machinery that targets myristoylation of proteins. Fungal cell wall target has been exploited for the development of caspofugin. Other drugs which are in experimental or clinical trials include the ones that target fungal components, e.g., nikkomycins, pradamicins, allylamines, sordarins, and cationic peptides [60]. A more recent approach to drug discovery is to utilize the genomics to search for targets that offers specificity and minimal toxicity. This would lead to improved strategies, diagnostic tools, and treatment programs for combating drug resistance. Moreover, recent sequencing and assembly of complete genome of *C. albicans* has permitted description of 6354 genes which can targeted for antifungal therapy [61].

SPREAD OF DRUG RESISTANCE

The dimorphic fungi *C. albicans* is able to assume different growth forms, which appear to have diverse roles for host–cell interaction pathogenicity and virulence [62]. A true hyphal form of *Candida* which may be involved in anchoring within and penetrating tissues is induced only in body temperatures and in the presence of inducing agents, whereas while at lower temperatures, in the absence of inducers, a unicellular yeast form is favored [63].

Another major side of these human fungal pathogens is formation of biofilm.

A biofilm is formed by a consortium or dense network of yeast and filaments attached to a surface and embedded in a protective extracellular matrix (ECM) [64]. The ability to form biofilms, both in the environment and in a host, provides protection to the microorganism from adverse conditions, including resistance to antimicrobial compounds and this poses a major problem for the treatment of many fungal infections. Dental plaque is an example of a complex mixed species biofilm, and is considered to be a source for recurrent oral candidiasis infections [65]. Many fungi, including *C. albicans* and *C. glabrata*, can form biofilms. Biofilms can also impound drugs in the polymers of the matrix and thus nullify their inhibitory effects. Fungal biofilms can also be formed in vivo on medical devices and human tissues, where they exhibit resistance to many antifungal drugs. *C. albicans* biofilm cells differ from planktonic cells in their metabolism and gene expression, and key molecules and regulatory networks are involved in fungal biofilm formation. Enhanced ECM especially β-glucan synthesis during biofilm growth has been shown to prevent penetration of antifungal agents such as azole and polyene. It is believed that the echinocandins and lipid formulations of amphotericin B can penetrate biofilm better than amphotericin B deoxycholate and azoles [66,67]. Many of the genes involved in, and regulated during, biofilm formation, or in the activities of *C. albicans* cells in biofilms, also contribute to the biofilm's resistance to antimicrobial compounds and immune evasion [68]. *C. albicans* biofilm formation proceeds in three developmental phases:

  i. Early phase (0–11 h), involving adhesion of fungal cells to the substrate;

ii. Intermediate phase (12−30 h), during which the blastospores coaggregate and proliferate, forming communities while producing a carbohydrate-rich ECM;

iii. Maturation phase (31−72 h), in which the fungal cells are completely encased in a thick ECM [69,70].

PERSISTENCE OF DRUG RESISTANCE

International Data

The increase in infections due to *Candida* over the past decade is significant. This is particularly true for hospitalized patients where the rate of blood-stream infection (BSI) due to *Candida* spp. has increased drastically. The global prevalence of candidemia was reported to be 6.9 cases per 1000 patients. In a 7-year long study analyzing nosocomial BSI in hospitals in United States, *Candida*-based infections are fourth most common and accounts for 8−10% nosocomial infections. The annual incidence rates are 6 to 13.3 cases per 100,000 populations. In Europe these are reported to be in top 10 most frequently reported hospital acquired BSIs with reported annual incidence rates ranging from 1.9 to 4.8 cases per 100,000 population. Researchers have documented a 50% increase in hospitalization with candidemia from January 2000 to December 2005. Later on increase in incidents of drug resistance has also been extensively reported.

C. albicans is the predominant cause of invasive candidiasis in the majority of clinical settings, accounting for 50−70% of cases however, and the epidemiology of *Candida* infection has changed in recent years, with longitudinal studies reporting that a considerable proportion of patients are now infected with non-*albicans Candida* (NAC) species. NAC species of clinical importance include *C. glabrata*, *C. tropicalis*, *C. parapsilosis*, and *C. krusei* which, together with *C. albicans*, account for more than 90% of cases of invasive candidiasis. Other less frequently reported *Candida* species include *C. guilliermondii*, *C. lusitaniae*, *C. kefyr*, *C. famata*, *C. inconspicua*, *C. rugosa*, *C. dubliniensis*, and *C. norvegensis*.

According to the results of studies in Spain, candidemia incidence in 2008−2009 and 2009−2010 was 1.09 cases/1000 admissions and 0.92 cases/1000 admissions, respectively. In Italy, an increase in the candidemia rate was reported between 1999 and 2009 according to the results of multicenter studies. The rate was found as 1.19 cases/ 1000 admissions in 2009, while this rate was 0.38/1000 admissions in the 1997–1999 period. A study including seven countries of South America conducted between 2008 and 2010 reported a candidemia incidence of 1.18/1000 admissions. In the study, the country with the highest incidence rate was Argentina (1.95 cases per 1000 admissions), followed by Venezuela (1.72 cases per 1000 admissions), Brazil (1.38 cases per 1000 admissions), Honduras (0.90 cases per 1000 admissions), Ecuador (0.90 cases per 1000 admissions), and Chile (0.33 cases per 1000 admissions) [71].

A 13-year-long study on candidemia from a tertiary care hospital in Thailand showed a prevalence of 6.14% for *Candida* species among blood culture isolates. Among the NAC species, *C. glabrata* has emerged as an important opportunistic pathogen worldwide. It is the second most common yeast isolated as part of normal flora and its role as a pathogen has only been recognized in the past few decades [72].

National Data

Candidemia is a life-threatening fungal infection associated with a mortality rate of 38%. There has been a lot of variation in the prevalence and incidence reports quoted from different parts of India. A study by Verma et al. from SGPGI in Lucknow ranked *Candida* species as eighth among all isolates from BSI. This study reported an incidence rate of 1.61 per 1000 hospital admissions for candidemia. A New Delhi-based study gave a prevalence rate of 18% for *Candida* species among blood culture isolates. A study in South India reported an incidence rate of 5.7% for candidemia among children with onco-hematological malignancies. Another study from Rohtak, North India, reported an isolation rate of 8.1% for *Candida* species from cases of neonatal septicemia. Xess et al. from AIIMS, New Delhi, found a prevalence rate of 6% for *Candida* species in a 5-year study (2001–2005). A study in Maulana Azad Medical College, New Delhi, found an incidence rate of 6.9% for *Candida* species in BSI. *C. krusei accounts for about 2–4% of all BSI caused by Candida species and are* especially important in patients with hematological malignancies and bone marrow transplants. *C. dubliniensis* is commonly misidenti-fied as *C. albicans* because of similar phenotypic characteristics like production of chlamydospores. This species has been isolated from a few cases of candidemia in the recent years [43].

The current retrospective analysis of candidemia over a 10-year period revealed a fivefold increase center. Between 2006 and 2008, *C. tropicalis* was the most common species (182 cases; 29.2%), followed by

C. albicans (105 cases; 16.8%) and
C. haemulonii (97 cases; 15.5%)
C. parapsilosis (78 cases; 12.5%)
C. glabrata (53 cases; 8.5%)
C. krusei (11 cases),
C. pelliculosa (23 cases)

These six species constituted 82.6% of the isolates [73].

MULTIDRUG RESISTANCE

MDR is defined as the resistance of an organism towards a number of structurally and functionally unrelated compounds and is the result of a synergistic action of a number of mechanisms. It is a complicated multifactor phenomenon. Pathogens with cross-resistance possess one mechanism that provides the ability to withstand more than one drug with different structural classes and families where as fungus with multiple drug resistances possess more than one mechanism that provides it the ability to tolerate most of the drugs.

For instance, R6G is a dye known to be effluxed from the cell by the pumps Cdr1p and Cdr2p. R6G has been shown to be capable of competing for FLC efflux. Not only R6G but 5-FC also has been suggested to be co-transported with FLC. A resistance to FLC automatically renders tolerance towards the ones that are co-transported [74,75].

CHAPTER 5

Multidrug Resistance and Transporters

Azole antifungals are most extensively used for treatment of infections and this has led to the development of resistance in yeast. The usage of drugs invariably also renders resistance. This acquired azole resistance results in cross-resistance to many unrelated drugs in the pathogen. This phenomenon is called multidrug resistance (MDR). MDR is often caused by upregulation of drug efflux pumps. *Candida* beats intercellular drug accumulation using efflux pumps, in particular ABC transporters and transporters of the MFS. These two most extensively studied fungal superfamilies of transporters play significant role in MDR to antifungal agents.

OVERVIEW OF EFFLUX PUMPS

ABC Transporters

The ABC proteins are primary transporters that are characterized by the presence of an ABC. ABC transporters are the most diverse and largest superfamily ubiquitously present in both prokaryotes and eukaryotes. The majorities of ABC transporters are membrane bound and contain two distinctive protein domains: nucleotide-binding domains (NBDs) and transmembrane domains (TMDs), i.e., $(NBD-TMD)_2$ topology [76].

TMDs are thought to create a path for solute movement across the phospholipid bilayer. Resistance of cells to chemotherapeutic agents and MDR in infectious microorganisms often arises from the overexpression of ABC transporters. Cdr1p (*Candida* drug resistance 1 protein) was the first ABC transporter identified as a drug efflux pump in *C. albicans* and shows a close homology (26%) to the Human ABC transporter P-glycoprotein (P-gp/MDR1).

Along with Cdr1p, Cdr2p has also been identified as a drug efflux pump and they share a close homology to each other. Both these pumps have been shown to efflux a wide variety of structurally

Anticandidal Agents. DOI: http://dx.doi.org/10.1016/B978-0-12-811311-0.00005-3

unrelated substrates including drugs, though Cdr1p has a wider range of substrate efflux as compared to Cdr2p even though they share such a high homology (91.8%) [77]. The molecular mechanisms that govern the function of Cdr1p or Cdr2p as efflux pumps for drugs are not well known.

The *Candida* database predicts several ABC transporters which show NBD conserved sequences upon multiple alignments. Although many transporters are predicted to have an (NBD-TMD)$_2$ topology, there are also some putative ORFs that have only (NBD-TMD)$_1$ topology and thus appear to be half proteins. Several ABC transporter genes (*CDR1, CDR2, CDR3, CDR4,* CDR5, *CDR11, SNQ2, YOR1*) of *C. albicans* were shown to be upregulated in the presence of azoles *CDR1* and *CDR2* has functional transporter of known antifungal drugs in *C. albicans* [78]. Disruption of *CDR1* makes *C. albicans* hypersensitive to azoles, and controlled overexpression of Cdr1p in *C. albicans CDR1*-null mutant conferred resistance to fluconazole (FLU) and other xenobiotics [76].

Among all the putative ABC proteins, only *CDR1* and *CDR2* have been experimentally implicated in *Candida* drug resistance.

MFS Transporters

The MFS consists of membrane transport proteins found from bacteria to higher eukaryotes and involved in antiport, symport, or uniport of various substrates. Antiporters transport two or more substrates in opposite directions across the membrane. Symporters translocate two or more substrates in the same direction simultaneously, making use of the electrochemical gradient of one substrate as the driving force; and uniporters transport only one type of substrate and are energized solely by the substrate gradient. The MFS pumps are secondary transporters that utilize the PMF across the plasma membrane. It is smaller in size compared to the ABC transporters since they do not require an NBD; however, almost all MFS proteins possess a uniform topology of 12 transmembrane α-helices connected by hydrophilic loops, with both their N and C termini located in the cytoplasm [79]. *CaMDR1* (*C. albicans* MDR 1) encodes a MFS protein in *C. albicans* whose expression has been linked to azole resistance. Some other MFS transporter genes (*FLU1, TPO3, orf19.2350, NAG3,* and *MDR97*) are involved in drug efflux. Mdr1p and Flu1p were shown in FLU hypersusceptibility for the development of azole resistance in clinical isolates of *C. albicans*. Disruption of *MDR1* in *C. albicans* resulted in reduced FLU susceptibility, and *MDR1* was shown to be upregulated in some *C. albicans* strains with reduced FLU susceptibility.

Previous study showed that the transporters Cdr1p, Cdr2p, and Mdr1p are the main efflux pumps mediating resistance of *C. albicans* to drugs. However, Mdr1p is relatively specific for FLU, whereas many azole drugs can act as substrates for Cdr1p and Cdr2p. The expression study of *CDR1* protein in *C. albicans* clinical isolates with reduced FLU susceptibilities demonstrated that Cdr1p was expressed in greater amounts than Cdr2p and that most FLU efflux function in these strains was mediated by Cdr1p rather than Cdr2p [80]. *CDR1*, *CDR2*, and *MDR1* are the most prominent contributors of MDR in *C. albicans* because most of the clinically drug-resistant isolates of *C. albicans* are shown to over express the genes encoding *CDR1*, *CDR2*, or *MDR1* proteins. MDR in *Candida* is due to the altered expression of drug efflux pump and the redundant nature of the transporters creates difficulty in characterization of the efflux mechanisms.

High levels of azole resistance in fungi are conferred by specialized efflux pumps which are encoded by two main protein families, ABC group and the MFS. ABCs are primary transporters that transport structurally unrelated drugs by using the energy gained from ATP. MFS pumps are secondary transporters that utilize the PMF and electrochemical gradient across the membrane to translocate substrates. ABC pumps comprises of both nucleotide-binding domains (two NBDs) and transmembrane domains (two TMDs) whereas MFS pumps have only TMD. These provide substrate specificity and transport through ABC is carried out at the NBD. The NBDs are involved in ATP binding and hydrolysis, and the TMDs span the membrane usually six times, via putative α-helices. The arrangement of the NBDs and TMDs within the pump polypeptide varies according to the type of ABC protein. "Full-size" fungal ABC transporters contain two NBDs and two TMDs and consist of about 1500 amino acids with a molecular mass of approximately 170 kDa. Fungi also contain "half-size" transporters with only one NBD and one TMD (Fig. 5.1). These half-size transporters may need to form homodimers or heterodimers to transport substrates.

MFS transporters, like ABC transporters, comprise large superfamilies of proteins with high sequence similarity found in plants, animals, bacteria, and fungi. There are two subfamilies of MFS transporters involved in drug efflux that are defined by the number of transmembrane spans (TMS) within the TMD:DHA1 (drug:H^+ antiporter 1; 12 TMS) and DHA2 (14 TMS). The first MFS transporter gene to be characterized from a pathogenic fungus was *CaMDR1* (also named *BEN^r*) [80].

A total of 28 ABC proteins in *Candida* have been subdivided in to six subfamilies. These can be classified as belonging to the following subclasses: PDR (pleiotropic drug resistance), MDR, MRP (multidrug resistance-associated protein), ALDp (adrenoleukodystrophy protein), YEF3 (yeast elongation factor EF-3), RLI (RNAse L inhibitor), and two that do not fit into the standard subclasses. Some of these proteins, although they possess NBDs, lack TMDs. These proteins are likely to be cytoplasmic and not involved in transport. The ABC proteins most frequently associated with fungal azole resistance are those belonging to the PDR, MDR, or MRP subclasses. These efflux pumps were named *Candida* drug resistance (CDR). Out of all these ABC proteins, seven have confirmed MDR which are CDR 1, 2, 3, 4, 5(11), SNQ2, and YOR1. Six genes are annotated as MFS but only CaMdr1 protein and CaFLu1 have been involved in antifungal resistance. Researches have proved that the ABC proteins Cdr1 and Cdr2 have been shown to contribute more frequently to the azole resistance than the MFS transporter Mdr1 [76].

The crux of the story is that two MFS genes in *C. albicans* are *CaMDR1* and *FLU1*. *CaMDR1* was initially identified as a gene that conferred resistance to benomyl which is the tubulin-binding agent and to methotrexate, a tetrahydrofolate reductase inhibitor. *FLU1* was initially isolated as a clone, which conferred resistance to FLU. Recently mycophenolic acid has been shown to be its specific substrate.

The contribution of drug transporters in antifungal drug resistance in the non-*albicans* species like *C. glabrata* is worth mentioning. *CgCDR1* and *CgCDR2* are upregulated in the azole-resistant *C. glabrata* isolates and have emerged as important nosocomial pathogen. A potential role of two putative ABC transporters ABC1 and ABC2 in drug resistance has also been suggested for *C. krusei*. The reduced accumulation of FLU in azole-resistant isolates of *C. dubliniensis* led to the identification of two ABC transporters CdCDR1 and CdCDR2 [81].

FUNGAL EFFLUX-MEDIATED DRUG RESISTANCE

Alteration of drug efflux is one of the prominent mechanisms of resistance in fungi. The *C. albicans CDR1* gene is a homolog of *S. cerevisiae* PDR5, which encodes a multidrug efflux pump, and *CDR1* is the gene most often associated with energy-dependent drug efflux in FLU-resistant clinical isolates [82]. The MDR phenotype in *C. albicans* has previously been shown to be linked to proteins encoded by *CDR1*, *CDR2*, and *MDR1* genes. These proteins act as membrane-localized efflux pumps that pump drugs from the fungal cells [83]. The ABC family is known to transport structurally unrelated drugs; this property has made a huge dent on the therapeutic research because once the resistance is attained for one class of agent, it is also acquired for all the other ones translocated along irrespective of its chemical class. CaCdr1 and CaCdr2 are main proteins involved in major efflux-mediated drug resistance [84].

OVERCOMING EFFLUX-MEDIATED DRUG RESISTANCE

In clinically used echinocandins and polyenes administered at therapeutic concentrations no major evidence of efflux-based resistance is reported suggesting that they are not pump substrates. Hence more and more drug bases that are not substrates of efflux pumps are needed. Few basic approaches to overcome the resistance would involve inhibiting the pumps, blocking the energy source of these pumps, and modulating transcriptional regulators of efflux pumps.

Inhibitors in combination with drugs directly act upon CaCdr1 and CaCdr2 blocking the primary transport completely (Fig. 5.1A). Since the gene expression pathway of both *CDR & MDR* genes is not dependent on each other, using inhibitors against MFS pumps would not necessarily inhibit ABC pumps. But it could be anticipated to reduce the rapid efflux of the drugs to a considerable level since the energy required for drug efflux is dependent on PMF (Fig. 5.1B). Targeting the transcription of genes *CDR 1* and *2* would not allow the synthesis of pump proteins take place. Similarly it would be interesting to see the results of transcriptionally inactivating *MDR* genes [85]. An overview of the above-mentioned approaches is given below (Fig. 5.1).

Adjuvants could be designed to surpass the efflux maintaining substantial concentrations of antifungals in the cell. Few agents like

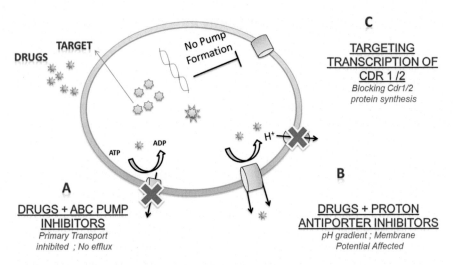

Figure 5.1 Plausible mechanisms to surpass efflux-mediated azole resistance.

nano-antibodies can be designed that recognize antigenic determinants on surface component of CDR and MDR. Drug designing with an enhanced rate of uptake and thus shift the balance between uptake and efflux so that a high intracellular concentration of the drug is maintained despite any upregulation of efflux [80].

CHAPTER 6

Potential Anticandidal Targets

Initially only a few contributing factors were considered to play a role in pathogenicity—(1) Release of enzymes of hydrolytic nature (extracellular), (2) Antigenic variability, and (3) Molecular mimicry. But with time new researches confirmed virulence dependent on dimorphic transition (filamentation) and ability to form biofilms. Sessile cells of biofilms display high levels of resistance. Virulence is not just based on multiple internal factors but it also depends on host immunity mechanisms and a peculiar fungus—host interaction.

SURFACE TARGETS

An attractive antifungal drug target is the fungal cell wall because the structure is absent from host cells and, thus, molecules that inhibit its synthesis are likely to have low human toxicity. Although the bacterial cell wall is the target of a number of archetypal antibiotics (e.g., penicillin), only one class of antifungal drugs, the echinocandins, targets the fungal cell wall [86]. They target β-1, 3-glucan which leads to faulty cell wall and defective hyphae in molds. Success of these drugs highlights the possible presence of other potential targets in vicinity. Similarly researches are carried out targeting β-1, 6-glucan to be a probable target. It forms strong network with chitin and β-1, 3-glucan where first-class proteins are linked to β-1, 6-glucan.

New leads pursuing processes critical to biosynthesis of cell wall can also be detected. Few scientific groups have reported certain molecules interfering with the biosynthesis of GPI. GPI-anchored proteins are important for homeostasis of membrane and cell wall. As mentioned earlier GPI proteins crosslink to β-1, 6-glucan creating network.

Also those elements which are unique to fungi are extensively researched. Other aspects like microtubule synthesis signal transduction and cell cycle could be exploited. Focusing on them increases the target

Anticandidal Agents. DOI: http://dx.doi.org/10.1016/B978-0-12-811311-0.00006-5

specificity as they are not shared with the host. As major drug resistance mechanisms are carried out by transporters of cell so these surface targets could be studied with further precision.

ABC Transporters

These proteins are primary active transporters that derive energy from the hydrolysis of ATP to drive the efflux of drugs. 28 ABC proteins can be grouped into five subfamilies. PDR, MDR, MRP, RLI/ALDP, and YEF3 (yeast elongation factor EF-3), and a sixth category that includes proteins unrelated to fungal subfamilies. Of the *C. albicans* ABC proteins containing both TMDs and NBDs, ten are full-size, three are full-size with an N-terminal TMD extension, five are half-size, and one is predicted to have two TMDs and one NBD, rest do not have a TMD. Further PDR subfamily consists of seven full-size transporters out of which only two are reported to have multidrug transport facility (Tables 6.1 and 6.2).

The MFS superfamily consists of 61 families according to the Transport Commission system given by Saier and group. Although, earlier annotation of the previous *Candida* genome assembly predicted 71 MFS genes, no systematic classification was given [87,88] (Table 6.3).

KEY FACTORS

In previous studies CDR2 mutants were found vulnerable to azoles and its overexpression conferred FLU resistance. But an enhanced hypersusceptibility was detected in CDR1 mutants. PDR subclass

Table 6.1 List of Proteins Involved in Drug Efflux Resistance Mechanism in *Candida*			
Protein	Family	Function	Location
Cdr1	ABC	Drug efflux	PM
Cdr2	ABC	Drug efflux	PM
Cdr3	ABC	Opaque phase-lipid translocation	PM
Cdr4	ABC	Flippase; translocation	PM
Cdr5 (11)	ABC	NC	NC
Snq2	ABC	NC	NC
4531	ABC	NC	NC
Mdr1	MFS	Drug efflux	PM
Flu1	MFS	Drug efflux	PM
PM, plasma membrane; NC, not characterized.			

Table 6.2 ABC Proteins: At a Glance

Class	Gene	Function
PDR	CDR1	Antifungal resistance
PDR	CDR2	Antifungal resistance
PDR	CDR3	Translocation
PDR	CDR4	Translocation
PDR	CDR5 (11)	NC
MDR	HST6	Biofilms
MDR	MDL1	Communication
MRP/CFTR	YOR1	Antifungal resistance
MRP/CFTR	YCF1	Transport glutathione conjugates
MRP/CFTR	MLT1	Sequester toxic from vacuoles
ALDp	–	NC
RL1	–	NC
EF3	ELF1, EF3	M-RNA export factor

Table 6.3 List of MFS Proteins

Gene Family	Genes	Function
DHA1	MDR1	Overexpression in resistant isolates, substrate for methoxtrate
DHA1	FLU1	Involved in efflux
DHA1	NAG3	Required for cycloheximide resistance
DHA1	NAG4	Required for cycloheximide resistance
DHA2	ORF19.2350	Affects filamentous growth
DHA2	ORF-CNC03290	Mutation causes increased azole sensitivity

DHA1, drug:H^+ antiporter (12 TMS); DHA2, drug:H^+ antiporter (14 TMS).

proteins cdr3, although shares a sequence homology with cdr1 and cdr 2 but no major role in drug resistance was found. Cdr4 may be involved in responses to osmotic and heavy metal stresses. But CDR 3 and 4 genes encode flippase and translocate membrane phospholipids. Function and localization of cdr5/cdr11 is still not characterized. Yor1 is also a plasma membrane-located *C. albicans* ABC transporter, of the MRP subclass, which is involved in resistance to aureobasidin A [89]. The ABC family in *C. albicans* comprises many genes, but apart from CDR1 and CDR2 no evidence for the involvement of other members in FLU resistance has been obtained so far; among these CDR 1 has a relatively major role [90].

MFS Transporters

These are secondary active transporters that utilize the electrochemical gradient of protons across the plasma membrane to efflux substrates out of the cell. The genome wide inventory revealed that *C. albicans* genome has 95 putative MFS proteins that clustered into 17 families. Out of all MFS proteins, only CaMdr1p is known to extrude drugs, where its overexpression has been linked to azole resistance [91].

FLU1 is another MFS gene (DHA1). Researches suggest that deletion of FLU1 has little effect on FLU as it was expected but there were evidences that suggest it being a pump substrate when the disruption resulted in cells being sensitive to immunosuppressant *Mycophenolic acid*. CaMdr1 protein is specific to FLU. Interestingly upregulation of CDR and MDR have two different pathways and are totally independent from each other. A lot of work has been done in the field of ABC expression and azole resistance as compared to MFS pumps [92].

QDR

The ABC proteins most commonly related with azole resistance belong to the subclasses: PDR, MDR, or MRP. Recently, *QDR1*, a putative quinidine drug resistance gene, was identified in genome expression profiling studies as a highly upregulated gene in *C. albicans* cells [93].

CELLULAR TARGETS

The improvement of already existing antifungal drugs and the limitation of drugs resistance apparition have helped to elucidate the basic biology of the fungal pathogen. An important difficulty in antifungal therapy is to develop drugs that exploit factors unique to fungi, which can be challenging considering that organism are eukaryotic and share many conserved biological pathways. Genes that are essential to fungal survival are possible targets for drug development [94]. For instance, TAC1 is localized in nucleus and it is the transcriptional activator of CDR genes. It was reported that TAC1 mutants expressed a lack of upregulation of CDR1 and CDR2. CAP1 is a transcription factor that negatively regulates MDR1 and was reported to be involved in drug resistance and oxidative stress response. FCR1 and FCR3 genes encode a transcription belonging to B-Zip family but no involvement in antifungal resistance is identified [86]. Gene encoding for transcription factor NDT80 also regulates expression of CDR genes. Interestingly, Ndt80 was crucial for the expression of ERG genes including ERG2, ERG25, ERG4, ERG24, ERG13, ERG9, ERG3, ERG10, ERG251, ERG1, ERG5, ERG6, ERG7, ERG11, and ERG26. Therefore, because Ndt80 modulates sterol metabolism and drug resistance in *C. albicans*, it represents a major element in the drug response of this yeast species.

Some ABC proteins are located primarily in organellar membranes such as Mlt1, located in the vacuolar membrane, and may confer resistance to metals by intracellular sequestration in the vacuole [82]. Firstly reported in the yeast *S. cerevisiae* target of rapamycin (TOR) is a cluster of proteins that are involved in activation and control of protein synthesis. *Mammalian translation of proteins involve EF-1 and EF-2 but fungi requires an additional factor called EF-3 for its effective translation. The uniqueness of this factor could be employed efficiently as it is* not present in humans. Genes of *C. albicans* in lipid, fatty acid, and sterol synthesis include ERG2, ERG3, ERG5, ERG10, ERG11, ERG25, and NCP1. Out of this ERG11 gene product is the target of azole antifungals. ERG3 encodes for C5 sterol desaturase. Mutations in ERG3 lowers ergosterol levels in the membrane and confers Amphotericin resistance. Genes PHR1, ECM21, ECM33, FEN12 code for proteins involved in maintenance for cell wall. The expression of these genes is affected by Caspofungin. The maintenance of cell calcium homeostasis is required for the survival and pathogenicity of fungi. Studies of calcium homeostasis systems and calcium signaling pathways indicate that they are closely associated with numerous physiological processes in *C. albicans*, such as stress responses, virulence, hyphal development, and adhesion. Secreted hydrolytic enzymes enable fungi to breach and invade host tissues. Recently secreted hydrolytic enzymes have attracted much attention as potential virulence factors in fungi. The most highly recognized extracellular hydrolytic enzymes are proteinases, lipolytic enzymes, lipases, and phospholipases. A number of studies suggest that the absence or decreased expression of these hydrolytic enzymes may lead to reduced virulence of *Candida* species, and these enzymes have been shown to contribute to *C. albicans* morphological transition, colonization, cytotoxicity, and penetration of host [95]. There are other possible pathogenic processes and potential targets in *C. albicans* that have not been investigated in detail that are intriguing, such as the metabolic pathway of arachidonic acid and ROS homeostasis. Prostaglandin production from arachidonic acid is critical for growth in *C. albicans* and could be a significant virulence factor in biofilm-associated infections, revealing great implications for understanding the mechanisms of *Candida* infections [96].

CROSS TALK

For improvement of antifungal therapy, cross talk involved within *C. albicans* in interaction of host and fungus may provide some prospective target leads. A major repertoire of proteins and other factors are involved during infection process.

Ndt80 transcriptional regulators include regulators of

- promoters of multidrug transporter genes (CDR1, CDR2, CDR4, and orf19.4531),
- hyphal growth (EFG1, NRG1, UME6, TEC1, CPH2, FLO8, CZF1, SSN6, RFG1),
- general transcriptional regulators (SUA71, TBP1, STP1, STP2P, STP3, STP4),
- carbohydrate metabolism (RGT1, TYE7, GAL4, MIG1),
- translation and amino acid metabolism (CBF1, GLN3, GCN4),
- lipid metabolism (INO2, OPI1, CTF1),
- cell cycle (SWI4, ASH1),
- stress (CAT8, HAC1, CAS5).

Ndt80 also attaches itself to promoter regions of MFS drug transporters such as MDR1 and FLU1. The transcription factor involved in MDR1 upregulation in a clinical strain was called multidrug resistance regulator 1 (MRR1). MRR1 inactivation in azole-resistant isolates resulted in the loss of MDR1 expression and increased susceptibility to FLU. Scientists have demonstrated successfully that MRR1 regulates MDR1. Although MRR1 has not been shown to bind directly to the MDR1 promoter as yet, it is likely that this transcription factor binds directly or indirectly the regions identified as BRE or MDRE. MDR1 can also be controlled by regulator of efflux pump 1 (REP1), which belongs to the transcription factor family including NDT80, was acting as a negative regulator of MDR1. Researchers suggest that when deleted in *C. albicans*, REP1 decreased azole susceptibility and also resulted in increased expression of MDR1 in the presence of an inducer. Interestingly, in the absence of both

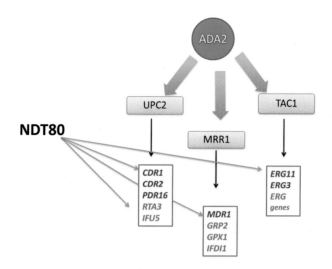

Figure 6.1 A cross talk between different proteins of Candida.
Genes mentioned in red color: Involved in antifungal resistance;
Genes mentioned in green: Involved in basic functioning;
Blue arrows represent the correlation between different genes and transcriptional factor NDT80.
Black arrows above depict the activation of cascade of genes by single gene.

REP1 and MRR1, MDR1 could still be upregulated in the presence of a drug. UPC2 is transcriptional activator of ERG11 and in absence of UPC2 failure of upregulation in ERG 11 is reported.

Ada2 is an adaptor protein that extensively takes part in gene regulation. It is recruited by TAC1, MRR1, and UPC2 that in turn regulates their specific set of dependent genes. Genes controlled by these major transcriptional activators exhibit a cross talk between their target genes. TAC1 and UPC2 regulate CDR1 and CAP1 regulates both PDR16 and MDR1, this is a perfect example of how genes regulate their subset and also how genes regulate genes found in separate regulons respectively.

Few regulatory factors can target other factors, which themselves are associated with other transcriptional units. Like EFG1, a regulator of morphogenesis is targeted by both CAP1 and NDT80 [97,98]. Even though the entire repertoire of interactions remains to be discovered, the existing data already suggest complex relationships [99,100] (Fig. 6.1).

CHAPTER 7

Drug Development Strategies

During the course of drug development numerous strategies have been taken up which could be further subcategorized as in vivo, in vitro, and in silico in their true forms. The benefits of in vitro tests lie in the fact that they render rapid results, relatively economical, and their precise modes of action can be tested. The drawback of these tests is that the homeostatic mechanisms and pathways found in animals are not present [101]. In vitro studies clearly signify prominent effects of an agent in a controlled environment outside of a living organism and hence confirm the credibility of a drug. But the confirmation through in vivo studies is better suited for its effects of an experiment on a living subject including the involved mechanisms of action. As compared to in vitro experiments, in vivo studies have greater probability of delivering substantial results of conclusive and deducible insights regarding nature of the drug and its effect on the host. Contrastingly in silico studies signify a relatively novel aspect of probing and analysis. Major conclusions drawn by implementing in silico approaches happen to be in the radar of prediction, modeling, and simulation regarding drug interaction within the host. While in silico and in vitro models will continually be developed and refined, in vivo preclinical safety models remain the gold standard for assessing human risk [102] (Fig. 7.1).

IN VITRO BASED

Biochemical Estimation

For establishing a correlation between drug dose, the minimum inhibitory concentration (MIC) for an organism, and outcome of the drug on the host body, both in vitro and in vivo models have proven to be successful. Studies carried out by Andes in the year 2003 demonstrated three patterns of antimicrobial activity that proves to be essential in understanding pharmacodynamic associations. Over the course of

Anticandidal Agents. DOI: http://dx.doi.org/10.1016/B978-0-12-811311-0.00007-7

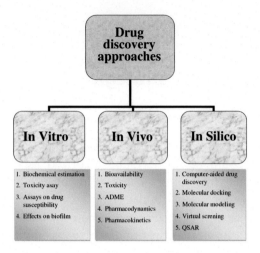

Figure 7.1 Classification of strategies implemented in the process of drug discovery and development.

antimicrobial activity their patterns have been depicted by probing the relationship between drug concentration and antimicrobial effect [103]. It was then established that the polyene amphotericin B and drugs of the new echinocandin class have been shown to exhibit concentration-dependent killing. When antimicrobial killing is enhanced by increasing drug levels, the pattern of activity is referred to as "concentration dependent." While some other drugs are referred to as "time dependent" where their effects are not altered by enhancing the concentrations but by increasing the duration of exposure. FLU and 5-FC are both time dependent whereas studies with drugs from the polyene and echinocandin class have demonstrated concentration-dependent killing [104].

Toxicity Assay

Researchers consider that in vitro toxicity analysis assays and techniques are relatively significant with respect to time and cost constrains as compared to in vivo toxicology studies in animal models. Cytotoxicity assays of drugs are carried out to analyze its relative toxicity to fungi but not to various human cell lines. A drug is studied for its candidal invasion of human cells and its effect is crucial not only in *C. albicans* but also in NAC species, such as *C. glabrata, C. tropicalis, C. krusei,* or *C. parapsilosis.* The in vitro antifungal activity of any novel drug is comparable to that of commonly used antifungal drugs such as amphotericin B or azoles. For that multiple in vitro assays are carried out. For instance 3-(4,5-dimethylthiazol-2-yl)-2,5-diphenyltetrazolium bromide (MTT) assay is performed to measure the cell viability. Moreover candidal adhesion assessment and biofilm development are carried out on surfaces like denture acrylic surfaces [103].

Another biochemical cytotoxicity analysis test is the MTS (3-(4, 5-dimethylthiazol-2-yl)-5-(3-carboxymethoxyphenyl)-2-(4-sulfophenyl)-2H-tetrazolium) assay. A modified MTT assay with a colorimetric basis has wee bit differences that make MTS method to be relatively better. Cell lysis based ATP assay renders results within minutes of performing and its need for relatively lesser samples makes it advantageous over other assays. Based on exhibited luminescence of cell's ATP content the number of live cells is deduced.

Drug Susceptibility

For studying drug susceptibility tests primarily MIC and minimum fungicidal concentrations (MFCs) are determined and pathogens' growth is monitored visually after incubation for 48 h at 308°C. The MIC was defined as the lowest concentration of drug showing no visible growth. The MFC was defined as the lowest concentration of drug which reduced the CFU by a definite value. In some cases agar diffusion assays on Yeast Nitrogen Base-glucose medium were also used to estimate drug susceptibility [105]. Another method to study the drug sensitivity was Fungitest method (Sanofi Diagnostics Pasteur). This is a microplate-based procedure for the breakpoint testing of antifungal drugs to analyze the sensitivity of *C. albicans* isolates against antifungals like FLU [106].

Effects on Biofilm

Different classes of drugs have been known to exhibit different effects on *Candida* biofilm over time. Some antifungals show better action against biofilms while some are ineffective due to resistance. Today azoles display no appreciable activity against both *C. albicans* and *C. parapsilosis* biofilms, whereas echinocandins like caspofungin and anidulafungin exhibit different results. The activity of anidulafungin and micafungin was compared with FLU, itraconazole, and voriconazole in an in vitro model against *C. albicans* and *C. parapsilosis* biofilms. In both studies, azoles showed less activity than the echinocandins against *Candida* biofilms [107]. For studying the effect on biofilms several studies have reported to perform XTT reduction bioassay [108]. Novel approaches like application of nanoparticles has been reported for the prevention and treatment of biofilm-associated candidiasis since biofilm formation by *C. albicans* complicates treatment with current available drugs [109].

IN VIVO BASED

Duration range of in vivo studies lies from short-term dosing to lifetime exposure. Short-term, acute studies are usually conducted in one or more rodent species like rat mice or gerbil with the sole aim to decide the test drug's lethality. Moreover dose range and the shape of the dose–response curve are also studied with in vivo models.

Andes et al in 2010 performed in vivo drug studies with infected neutropenic mice. The applicability of the infection model was studied in response to caspofungin, a common echinocandin drug used against *Candida* infections. To compare the in vivo potencies of drug against the various *Candida* strains and species, they utilized the dose level required to produce a net static effect. The dose response results were analyzed and it was established that caspofungin displayed variable potencies against different fungal species [110].

Similarly in 2008 they gave the study with micafungin against *C. albicans* and *C. glabrata* and results depicted that drug micafungin exposures required to produce stasis and killing endpoints were similar to those recently reported for another echinocandin, anidulafungin, against the identical *Candida* isolates in this model [111].

Another study involving *Candida* demonstrates that anidulafungin has concentration-dependent in vivo efficacy against both *C. albicans* and *C. glabrata*. The in vitro potency of caspofungin, micafungin, and anidulafungin, against *C. albicans* and *C. glabrata* are similar. In this study the pharmacokinetics of anidulafungin in animal models demonstrate that the drug has a long half-life and protein binding values essentially the same as those observed in mice [112]. Wong et al. have elaborated in vivo drug susceptibility studies on mouse model of systemic candidiasis as well as oral candidiasis [103].

Bioavailability

Determining the absolute bioavailability (BA) of a drug is done through a pharmacokinetic study in clinical trials precisely in Phase 1 and Phase 2 trials. BA refers to the degree and rate of absorption of the drug administered to human, the analysis of how and at what pace an administered drug is absorbed by the body's systemic circulation. To determine the correct amount for nonintravenous administration dose of a drug like oral, rectal, transdermal, subcutaneous, and sublingual, this parameter is essential [113]. Route of administration, drug metabolism, and the formulation of drug effect the absorption rate. Apart from these two primal cases BA depends on several factors; for instance, during the case of intravenous administration, the BA percentage (F) remains to be 100. But in oral administrations the rate is severely affected due to absorption by intestine before reaching circulatory system. To measure the absorbed amount of bioavailable drug in the host body there are perpetually two methods. One of them involves studying the area under the time-plasma concentration curve (AUC). The other is to determine the cumulative amount of drug excreted in the urine following drug administration [114].

Toxicity

Understanding the toxicity of drugs is crucial because majority of drugs fail due to its toxicity and safety issues. Toxicity studies provide a thorough investigation of the dose effect of a drug on various points of host. They are carried in the name chronic bioassay which is conducted in rodent species in both sexes, and for duration that approaches the lifespan of the animal [101]. In vivo virulence analysis of echinocandin-resistant mutant tested in a disseminated-candidiasis target organ assay with female mice models gave broader insights in the in vivo toxicity of a drug [105].

ADME
Absorption and Distribution

To analyze disposition of drugs upon entering the body and its kinetics upon exposure to tissues are studied by parameters like adsorption distribution metabolism and excretion. Major antifungal drugs like polyene and echinocandin don't exhibit significant BA when administered orally. For other classes the degree of absorption and optimal administration conditions vary extensively. Today, each member of the azole class can be administered orally; however, their extent of absorption varies. Oral BA of FLU is 90% thus termed as readily absorbed. Similarly voriconazole with oral BA easily achieves concentrations more than 90%. Sometimes antifungals from same classes exhibit different levels of appreciable BA depending upon the formulation and external conditions like surrounding pH [94]. Postabsorption the drug is distributed to the tissues via circulatory system.

Metabolism and Excretion
Many antifungal drugs exhibit hepatic metabolism during the whole process of elimination. For instance, drugs under the class azoles undergo different levels of metabolism whereas 5-FC is an exception in this case. Similarly echinocandin class of antifungal drugs exhibit hepatic hydrolysis. An exception of hepatic metabolism in this class is anidulafungin [94]. The last step where drug and its metabolites are removed is excretion. Elimination of drug and its metabolites takes place via filtration by kidney that results into urine or through biliary excretion resulting into feces.

Pharmacokinetics and Pharmacodynamics
In simplest terms the underlying difference between these two broader terms "Pharmacokinetics" and "Pharmacodynamics" (PD) are complementary. Former is what changes are transpired upon drug by the body whereas latter is what a drug does to body.

Pharmacokinetic methods are further classified into direct and relatively easy yet expensive indirect methods. Direct methods involve concentrations of drug measured invariably from tissue fluids directly. On the other hand indirect methods measure tissue penetrance and distribution with the aid of radioisotope labeled on drugs [115]. Pharmacokinetic studies in mice are carried out using BALB/c or CD-1 mice. Drugs administered intravenously from the tail vein or intraperitoneally at a fixed dose and pharmacokinetic modeling and parameter calculations were carried out using the WinNonlin pharmacokinetics software [116]. After pharmacokinetic studies, associations between drug and PD parameters are established. To determine the relation between drug exposure and treatment, essential parameters governing the pharmacodynamic studies are as follows:

a. Time > MIC where time is percentage of duration
b. Cmax/MIC where Cmax is the area under the serum concentration curve
c. AUC/MIC ratio [117]

IN SILICO BASED

Ligand-based drug design is rather indirect approach where analysis is totally dependent on information of other molecules that attach to the target compound of interest. These other molecules may be used to develop a pharmacophore model that delineates the minimum needed structural characteristics a molecule must hold in order to bind to the target successfully. Precisely a model of the biological target may be constructed on the facts of what binds to it. This knowledge could be further implemented to design new molecular entities that interact with the target.

Structure-based drug design which is relatively a direct approach is based on the knowledge of the three-dimensional structure of the biological target obtained through methods such as X-ray crystallography or NMR spectroscopy which are mainly the starting point for gathering information in the field of drug discovery and development. In case of unavailability of the target structure, prediction using a related protein's experimental structure is carried out by generating a homology model of the target [111]. Based on this data drug that seems to be exhibiting high affinity for the target is selected and further computational trials are employed to suggest new drug candidates.

Computer-Aided Drug Discovery and Development

Computer-aided drug design uses computational approaches to discover, develop, and analyze drugs and similar biologically active molecules. The ligand-based computer-aided drug discovery (LB-CADD) approach involves the analysis of ligands known to interact with a target of interest. These methods use a set of reference structures collected from compounds known to interact with the target of interest and analyze their 2D or 3D structures. The basic objective of these methods is to predict the nature and strength of binding of given molecule a target. Ab initio quantum chemistry methods, or density functional theory, are frequently used to deliver optimized parameters for the molecular mechanics calculations to predict the conformation of the small molecule and to model conformational changes in the biological target that may occur when the small molecule binds to it. The data also provide an estimate of the electronic properties (electrostatic potential, polarizability, etc.) of the drug candidate that will influence

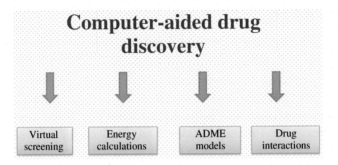

Figure 7.2 Characteristic features of CADD.

binding affinity [118]. CADD methods can surge the probabilities of recognizing compounds with desirable characteristics, hustle up the hit-to-lead development, and expand the odds of getting a compound over the many obstacles of preclinical testing (Fig. 7.2).

Molecular Docking

Docking is a pharmacologically important tool in the field of drug designing and computational biology. It works with the basic understanding of structure prediction of intermolecular complex formed between drug and its target molecule. It also gives other important information like extent and specificity of interaction, binding and transformation energies. The aim of ligand-receptor docking is to identify the pivotal active binding sites of a ligand with a protein of already known three-dimensional structure. This method is relatively new in the field of lead designing and analysis of drug−ligand interactions [119]. Modes of docking include rigid receptor-ligand flexible and receptor-ligand rigid. A lot of novel antifungal compounds that have been tested and further proceeded into trials with the aid of this technique [120]. Docking studies have been implied to study novel triazole derivatives and their interaction with their binding site [121].

Molecular Modeling

An in silico approach with pivotal function in the field of drug design and drug discovery, molecular modeling is a method of comparing and analyzing molecular structures, 3D structure-based properties, and their interactions. For identification of potential drug candidates, molecular modeling has become an important technique. The process of data mining and data interpretation is often followed by modeling and predicting the properties of small lead compounds as probable drug candidates. Primary applicability of this technique generates data on receptor–ligand interaction, protein–protein interactions, 3D structure prediction, and based on this predicting biological function of related compounds [122]. In early 2000s this technique flourished and many antifungal drugs were studied for their significant biological activities. For instance, in 2008 amphotericin B was studied for its membrane activity for better understanding of its mechanism of action [121].

Virtual Screening
During the course of drug discovery process, computational screening of new lead compounds is termed as virtual screening that comes under. Fortunately, there are now a number of drugs whose development was profoundly predisposed by structure-based design. They are heavily based on computational and virtual screening strategies [123]. Virtual screening is the key methodology applied in drug discovery studies for the identification of new hits. Receptor structure based as well as nonreceptor based methods could be employed to design ligand.

Quantitative Structured Activity Relationship
Quantitative structured activity relationship (QSAR) establishes a relation between predictor and response variables by applying graphical and mathematical models. For analysis the QSAR response-variable is the biological activity of the chemicals and the predictors consist of properties of chemicals. For prediction and evaluation of compound-binding modes mathematical relationships between structural attributes and target properties of a set of chemicals are studied [122]. The general workflow of a QSAR-based drug discovery project is to first collect a group of active and inactive ligands and then create a set of mathematical descriptors that describe the physicochemical and structural properties of those compounds. A model is then generated to identify the relationship between those descriptors and their experimental activity, maximizing the predictive power. Finally, the model is applied to predict activity for a library of test compounds that were encoded with the same descriptors. QSAR along with QSPR that stands for quantitative structured property relationship are mainly applicable in computational predictive toxicology. Multidimensional approaches like 3D, 4D, and 5D QSAR are also quite significant in the evaluation of drug bases and targets. 4D-QSAR is an extension of 3D-QSAR that treats each molecule as an ensemble of different conformations, orientations, tautomers, stereoisomers, and protonation states [124].

CHAPTER 8

Conclusion

Development of new antifungal therapeutics is essential. Unfortunately, the antifungal pipeline is dry and declined over the time. As compared to other antibiotics that have displayed a potent development with time. There is an ardent need for both prophylactic and therapeutic treatments worldwide to control the holistic causes of *Candida* overgrowth. There have been significant changes in the treatments that are used to manage *Candida* infections over time. Mostly there are several factors that one must keep in mind whilst working on drug discovery. Primarily beginning with cause and severity of disease, further on the site of infection, immune capacity, the risk factors, and susceptibility of pathogen to specific drugs should be profoundly considered.

Mostly the new triazoles that were recently formed were nothing but new versions of older models that do not address the emergence of resistance. With an already constricted list of antifungal drugs, alternate drug choices turn out to be tricky if resistance towards triazoles develops during therapy. We believe that there is a great need because the rise and expansion of resistance renders a significant challenge amongst researchers. Drug resistance in *C. albicans* occurs at a specific rate whilst the reason for such an event is totally dependent on the localization and functioning of specific drug target. Today drug resistance is a major setback in the field of current drugs that makes the process of drug discovery and development a scientific necessity.

Antifungal resistance is a crisis implying complicated hindrance in treating major life-threatening diseases. In pathogenic fungi scenario of MDR is an emergent area wherein new contributors of drug resistance are being documented with a massive stride. While some of the mechanisms of drug resistance are completely understood, a majority of them still require a closer look. The scene is still incomplete because

Anticandidal Agents. DOI: http://dx.doi.org/10.1016/B978-0-12-811311-0.00008-9

primarily it is attributed to interplay between multiple mechanisms. Understanding those underlying mechanisms of resistance helps build a platform for novel drug candidate search. Mechanism and structure-based approach of drug designing gives a stable foundation to the tardy process of drug discovery and development. Protecting the current antifungal drug pipeline as well as expanding it by incorporating novel inhibitors is vital.

ABBREVIATIONS

5-FC	5-fluorocytosine
5-FU	5-fluorouracil
ABC	ATP-binding cassette
AUC	area under curve
BQM	bis [1, 6quinolizinium 8-methyl-salt]
CADD	computer-aided drug designing
CDR	*Candida* drug resistance
CLR	C-type lectin receptors
CWP	cell wall protein
DAMP	damage-associated molecular patterns
DC	dendritic cells
DHA	Drug & H+ antiporter
dTMP	deoxythyminemonophosphate
dUMP	deoxyuracilmonophoshate
ECM	extracellular matrix
FLU	fluconazole
IFI	invasive fungal infections
IFN	interferon
IL	interleukin
INOS	inducible nitric oxide synthase
LBDD	ligand-based drug designing
MDR	multidrug resistance
MFS	major facilitator superfamily
MIC	minimum inhibitory concentration
MLC	minimum lethal concentration
NAC	non *albicans* Candida
NBD	nucleotide-binding domain
NLR	NOD-like receptors
ORF	open reading frame
PAMP	pathogen-associated molecular patterns
PD	pharmacodynamics
PDR	pleiotropic drug
PK	pharmacokinetics
PM	plasma membrane

PMF	proton motive force
PRR	pattern recognition receptors
ROS	reactive oxygen species
RVVC	recurrent vulvovaginal candidiasis
SBDD	structure-based drug designing
TLR	Toll-like receptor
TMD	transmembrane domains
TOR	target of rapamycin
XTT	(2,3-Bis-(2-methoxy-4-nitro-5-sulfophenyl)-2H-tetrazolium-5-carboxanilide)

BIBLIOGRAPHY

[1] F.C. Odds, "Candida and candidosis: a review and bibliography. 2nd edition.," 1988.

[2] C.O. Truss, "The Role of Candida Albicans in Human Illness," 2000.

[3] L. Ferna, A.B. Herrero, M.C. Lo, and A. Domı, "Candida albicans and Yarrowia lipolytica as alternative models for analysing budding patterns and germ tube formation in dimorphic fungi," pp. 2727−2737, 1999.

[4] W.E. Dismukes, Introduction to antifungal drugs, Clin. Infect. Dis. 30 (4) (2000) 653−657.

[5] N.H. Georgopapadakou, T.J. Walsh, Antifungal Agents: Chemoterapeutic Targets and Immunologic Strategies, Antimicrob. Agents Chemother. 40 (2) (1996) 279−291.

[6] Y.-L. Yang, Virulence factors of Candida species, J. Microbiol. Immunol. Infect. 36 (4) (2003) 223−228.

[7] C.J. Nobile, E.P. Fox, J.E. Nett, T.R. Sorrells, Q.M. Mitrovich, A.D. Hernday, et al., A recently evolved transcriptional network controls biofilm development in Candida albicans, Cell 148 (1−2) (2012) 126−138.

[8] C.A. Kumamoto, Molecular mechanisms of mechanosensing and their roles in fungal contact sensing, Nat. Rev. Microbiol. 6 (9) (2008) 667−673.

[9] F.L. Mayer, D. Wilson, B. Hube, Candida albicans pathogenicity mechanisms, Virulence 4 (2) (2013) 119−128.

[10] M.A. Kabir, M.A. Hussain, Z. Ahmad, Candida albicans: A Model Organism for Studying Fungal Pathogens, Int. Sch. Res. Netw. Microbiol. 2012 (2012) 1−15.

[11] "Candida: Comparative and Functional Genomics | Book." [Online]. Available from <http://www.horizonpress.com/can>. (accessed 11.09.15).

[12] F.C. Odds, A.J.P. Brown, N.A.R. Gow, Candida albicans genome sequence: a platform for genomics in the absence of genetics, Genome. Biol. 5 (7) (2004) 230.

[13] T.L. Han, R.D. Cannon, S.G. Villas-Boas, Metabolome analysis during the morphological transition of *Candida albicans*, Metabolomics. 8 (6) (2012) 1204−1217.

[14] Na.R. Gow, F.L. van de Veerdonk, A.J.P. Brown, M.G. Netea, P. Vandeputte, S. Ferrari, et al., Candida albicans morphogenesis and host defence: discriminating invasion from colonization, Int. J. Microbiol. 2012 (2) (2011) 112−122.

[15] G. Molero, F. Navarro-garcía, and M. Sánchez-pérez, "Candida albicans: genetics, dimorphism and pathogenicity," pp. 95−106, 1998.

[16] L. Romani, Immunity to fungal infections, Nat. Rev. Immunol. 11 (4) (2011) 275−288.

[17] M.V. Elorza, E. Valent, Molecular organization of the cell wall of Candida albicans and its relation to pathogenicity, FEMS Yeast Res. 6 (2006) 14−29.

[18] M. Whiteway, C. Bachewich, Morphogenesis in Candida albicans*, Annu. Rev. Microbiol. 61 (1) (2007) 529−553.

[19] C.G.J. Mckenzie, U. Koser, L.E. Lewis, J.M. Bain, R.N. Barker, N.A.R. Gow, et al., Contribution of Candida albicans cell wall components to recognition by and escape from murine macrophages, Infect. Immun. 78 (2010) 1650−1658.

[20] A. Butts, D.J. Krysan, Antifungal drug discovery: something old and something new, PLoS Pathog. 8 (9) (2012) 9–11.

[21] M.A. Ghannoum, L.B. Rice, Antifungal agents: mode of action, mechanisms of resistance, and correlation of these mechanisms with bacterial resistance, Clin. Microbiol. Rev. 12 (4) (1999) 501–517.

[22] K. Maebashi, M. Niimi, M. Kudoh, F.J. Fischer, K. Makimura, K. Niimi, et al., JAC mechanisms of fluconazole resistance in Candida albicans isolates from Japanese AIDS patients, J. Antimicrob. Chemother. 47 (2001) 527–536.

[23] N.N. Mishra, T. Prasad, N. Sharma, A. Payasi, R. Prasad, D.K. Gupta, et al., Pathogenicity and drug resistance in candida albicans and other yeast species, Acta Microbiol. Immunol. Hungarica. 54 (3) (2007) 201–235.

[24] J. Loeffler, Da Stevens, Antifungal drug resistance, Clin. Infect. Dis. 36 (Suppl. 1) (2003) S31–S41.

[25] Y.M. Welscher, H.H. Napel, M. Masia, C.M. Souza, H. Riezman, B. De Kruijff, et al., Natamycin blocks fungal growth by binding specifically to ergosterol without permeabilizing the membrane, J. Biol. Chem. 283 (2008) 6393–6401.

[26] R.A. Calderone, W.A. Fonzi, W.A. Fonzi, Virulence factors of Candida albicans, Trends Microbiol. 9 (2001) 327–335.

[27] Y. Li, L. Chen, Q. Yue, X. Liu, Z. An, G.F. Bills, Genetic manipulation of the pneumocandin biosynthetic pathway for generation of analogues and evaluation of their antifungal activity, ACS Chem. Biol. 10 (7) (2015) 1702–1710.

[28] D.W. Denning, Echinocandins: a new class of antifungal, J. Antimicrob. Chemother. 49 (6) (2002) 889–891.

[29] G.S. Hall, C. Myles, K.J. Pratt, J.A. Washington, Cilofungin (LY121019), an antifungal agent with specific activity against Candida albicans and Candida tropicalis, Antimicrob. Agents Chemother. 32 (9) (1988) 1331–1335.

[30] R. Pasrija, S. Krishnamurthy, T. Prasad, J.F. Ernst, R. Prasad, Squalene epoxidase encoded by ERG1 affects morphogenesis and drug susceptibilities of Candida albicans, J. Antimicrob. Chemother. 55 (6) (2005) 905–913.

[31] E.I. Mercer, Morpholine antifungals and their mode of action, Biochem. Soc. Trans. 19 (3) (1991) 788–793.

[32] C. Onyewu, J.R. Blankenship, M. Del Poeta, J. Heitman, Ergosterol biosynthesis inhibitors become fungicidal when combined with calcineurin inhibitors against Candida albicans, Candida glabrata, and Candida krusei, Antimicrob. Agents Chemother. 47 (2003) 956–964.

[33] P. Pasanen, A. Pasanen, K. Yli-pietila, Ergosterol content in various fungal species and bio-contaminated building materials, Appl. Environ. Microbiol. 65 (1999) 138–142.

[34] D.M. Dixon and T.J. Walsh, Antifungal Agents, University of Texas Medical Branch at Galveston, 1996.

[35] C. Jiménez-Ortigosa, P. Paderu, M.R. Motyl, D.S. Perlin, "Enfumafungin derivative MK-3118 shows increased in vitro potency against clinical echinocandin-resistant Candida species and Aspergillus species isolates", Antimicrob. Agents Chemother. 58 (2) (2014) 1248–1251.

[36] F. Lamoth, B.D. Alexander, Antifungal activities of SCY-078 (MK-3118) and standard antifungal agents against clinical non-Aspergillus mold isolates, Antimicrob. Agents Chemother. 59 (7) (2015) 4308–4311.

[37] Efficacy and safety of oral VT1161 a novel inhibitor of fungal CYP51 in a randomized phase 2 study (Online). Available from <http://www.viamet.com/sites/default/files/product-posters/2015-ECCMID_VT-1161_AVVC_Clinical.pdf>. (accessed 15.09.15).

[38] M.A. Pfaller, P.R. Rhomberg, S.A. Messer, M. Castanheira, "In vitro activity of a Hos2 deacetylase inhibitor, MGCD290, in combination with echinocandins against echinocandin-resistant Candida species", Diagn. Microbiol. Infect. Dis. 81 (4) (2015) 259–263.

[39] N.-A. Watanabe, M. Miyazaki, T. Horii, K. Sagane, K. Tsukahara, K. Hata, E1210, a new broad-spectrum antifungal, suppresses Candida albicans hyphal growth through inhibition of glycosylphosphatidylinositol biosynthesis, Antimicrob. Agents Chemother. 56 (2) (2012) 960–971.

[40] T. Shibata, T. Takahashi, E. Yamada, A. Kimura, H. Nishikawa, H. Hayakawa, et al., T-2307 causes collapse of mitochondrial membrane potential in yeast, Antimicrob. Agents Chemother. 56 (11) (2012) 5892–5897.

[41] H. Morikawa, M. Tomishima, N. Kayakiri, T. Araki, D. Barrett, S. Akamatsu, et al., Synthesis and antifungal activity of ASP9726, a novel echinocandin with potent Aspergillus hyphal growth inhibition, Bioorg. Med. Chem. Lett. 24 (4) (2014) 1172–1175.

[42] Biafungin (CD101), a novel echinocandin, displays a long half-life in the Chimpanzee (Online). Available from <http://www.cidara.com/wp-content/uploads/2014/12/A-694.-Biafungin-CD101-a-Novel-Echinocandin-Displays-a-Long-Half-life-in-the-Chimpanzee-Suggesting-a-Once-Weekly-IV-Dosing-Option.pdf>. (accessed 15.09.15).

[43] N. Sun, D. Li, W. Fonzi, X. Li, L. Zhang, R. Calderone, Multidrug-resistant transporter Mdr1p-mediated uptake of a novel antifungal compound, Antimicrob. Agents Chemother. 57 (12) (2013) 5931–5939.

[44] M.K. Kathiravan, A.B. Salake, A.S. Chothe, P.B. Dudhe, R.P. Watode, M.S. Mukta, et al., The biology and chemistry of antifungal agents: a review, Bioorganic. Med. Chem. 20 (19) (2012) 5678–5698.

[45] Q. Yu, C. Xiao, K. Zhang, C. Jia, X. Ding, B. Zhang, et al., The calcium channel blocker verapamil inhibits oxidative stress response in Candida albicans, Mycopathologia 177 (3–4) (2014) 167–177.

[46] A. Kumar, S. Dhamgaye, I.K. Maurya, A. Singh, M. Sharma, R. Prasad, "Curcumin targets cell wall integrity via calcineurin-mediated signaling in candida albicans", Antimicrob. Agents Chemother. 58 (1) (2014) 167–175.

[47] M. Sharma, R. Manoharlal, S. Shukla, N. Puri, T. Prasad, S.V. Ambudkar, et al., Curcumin modulates efflux mediated by yeast ABC multidrug transporters and is synergistic with antifungals, Antimicrob. Agents Chemother. 53 (8) (2009) 3256–3265.

[48] T. Prasad, P. Saini, N.A. Gaur, A. Ram, L.A. Khan, Q.M.R. Haq, et al., Functional analysis of CaIPT1, a sphingolipid biosynthetic gene involved in multidrug resistance and morphogenesis of Candida albicans, Antimicrob. Agents Chemother. 49 (8) (2005) 3442–3452.

[49] N.H. Ibrahim, N.A. Melake, A.M. Somily, A.S. Zakaria, M.M. Baddour, A.Z. Mahmoud, The effect of antifungal combination on transcripts of a subset of drug-resistance genes in clinical isolates of Candida species induced biofilms, Saudi Pharm. J. 23 (1) (2014) 55–66.

[50] P.D. Rogers, K.S. Barker, Evaluation of differential gene expression in fluconazole-susceptible and -resistant isolates of Candida albicans by cDNA microarray analysis, Society 46 (11) (2002) 3412–3417.

[51] L.A. Walker, D.M. MacCallum, G. Bertram, Na.R. Gow, F.C. Odds, A.J.P. Brown, Genome-wide analysis of Candida albicans gene expression patterns during infection of the mammalian kidney, Fungal Genet. Biol 46 (2) (2009) 210–219.

[52] L.E. Cowen, J.B. Anderson, L.M. Kohn, Evolution of drug resistance in Candida albicans, Annu. Rev. Microbiol. 56 (2002) 139–165.

[53] Ka Marr, C.N. Lyons, K. Ha, T.R. Rustad, T.C. White, Inducible azole resistance associated with a heterogeneous phenotype in Candida albicans, Antimicrob. Agents Chemother. 45 (1) (2001) 52–59.

[54] S. Hameed, Z. Fatima, Novel regulatory mechanisms of pathogenicity and virulence to combat MDR in Candida albicans, Int. J. Microbiol. 2013 (1) (2013).

[55] Advances in Applied Microbiology, Volume 84, Academic Press, 2013.

[56] A. Kohli, N.F.N. Smriti, K. Mukhopadhyay, A. Rattan, R. Prasad, In vitro low-level resistance to azoles in Candida albicans is associated with changes in membrane lipid fluidity and asymmetry, Antimicrob. Agents Chemother. 46 (2002) 1046–1052.

[57] T.L. Han, R.D. Cannon, S.G. Villas-Bôas, The metabolic basis of Candida albicans morphogenesis and quorum sensing, Fungal Genet. Biol. 48 (8) (2011) 747–763.

[58] K. Mukhopadhyay, A. Kohli, R. Prasad, Drug susceptibilities of yeast cells are affected by membrane lipid composition drug susceptibilities of yeast cells are affected by membrane lipid composition, Antimicrob. Agents Chemother. 46 (2002) 3695–3705.

[59] K. Mukhopadhyay, T. Prasad, P. Saini, J. Pucadyil, A. Chattopadhyay, R. Prasad, et al., Membrane sphingolipid-ergosterol interactions are important determinants of multidrug resistance in Candida albicans, Antimicrob. Agents Chemother. 48 (5) (2004) 1778–1787.

[60] N.N. Mishra, T. Prasad, N. Sharma, D.K. Gupta, Membrane fluidity and lipid composition of fluconazole resistant and susceptible strains of Candida albicans isolated from diabetic patients, Brazilian J. Microbiol. 39 (2) (2008) 219–225.

[61] G.P. Moran, D.C. Coleman, D.J. Sullivan, Candida albicans versus Candida dubliniensis: why is C. albicans more pathogenic?, Int. J. Microbiol. 2012 (2012).

[62] P.J. Brockert, S.A. Lachke, T. Srikantha, C. Pujol, R. Galask, D.R. Soll, et al., Phenotypic switching and mating type switching of Candida glabrata at sites of colonization, Infect. Immun. 71 (2003) 7109–7118.

[63] S. Krishnamurthy, A. Plaine, J. Albert, T. Prasad, R. Prasad, J.F. Ernst, Dosage-dependent functions of fatty acid desaturase Ole1p in growth and morphogenesis of Candida albicans, Microbiology 150 (Pt 6) (2004) 1991–2003.

[64] P.E. Sudbery, Growth of Candida albicans hyphae, Nat. Publ. Gr. 9 (10) (2011) 737–748.

[65] P. Mishra, J. Bolard, R. Prasad, Emerging role of lipids of Candida albicans, a pathogenic dimorphic yeast, Biochim. Biophys. Acta - Lipids Lipid Metab. 1127 (1) (1992) 1–14.

[66] M. Martchenko, A. Alarco, D. Harcus, M. Whiteway, Superoxide dismutases in Candida albicans: transcriptional regulation and functional characterization of the hyphal-induced sod5 gene, Mol. Biol. Cell 15 (2004) 456–467.

[67] A. Brand, Hyphal growth in human fungal pathogens and its role in virulence, Int. J. Microbiol. vol. 2012 (2012).

[68] N. Chauhan, D. Inglis, E. Roman, J. Pla, D. Li, J.A. Calera, et al., Candida albicans response regulator gene SSK1 regulates a subset of genes whose functions are associated with cell wall biosynthesis and adaptation to oxidative stress, Eukaryot. Cell 2 (2003) 1018–1024.

[69] L. Martinez, P. Falson, Multidrug resistance ATP-binding cassette membrane transporters as targets for improving oropharyngeal candidiasis treatment, Adv. Cell. Mol. Otolaryngol. 2 (2014) 1–8.

[70] P.E. Sudbery, The germ tubes of Candida albicans hyphae and pseudohyphae show different patterns of septin ring localization, Mol. Microbiol. 41 (2001) 19–31.

[71] N. Yapar, Epidemiology and risk factors for invasive candidiasis, Ther. Clin. Risk Manag. 10 (1) (2014) 95–105.

[72] C.M. Hull, J.E. Parker, O. Bader, M. Weig, U. Gross, A.G.S. Warrilow, et al., Facultative sterol uptake in an ergosterol-deficient clinical isolate of candida glabrata harboring a missense mutation in ERG11 and exhibiting cross-resistance to azoles and amphotericin B, Antimicrob. Agents Chemother. 56 (2012) 4223–4232.

[73] J.K. Oberoi, C. Wattal, N. Goel, R. Raveendran, S. Datta, K. Prasad, Non-albicans Candida species in blood stream infections in a tertiary care hospital at New Delhi, India, Indian J. Med. Res. 136 (6) (2012) 997−1003.

[74] B.E. Mansfield, H.N. Oltean, B.G. Oliver, S.J. Hoot, S.E. Leyde, T.C. White, Azole drugs are imported by facilitated diffusion in Candida albicans and other pathogenic fungi, PLoS Pathog. 6 (2010) 1−11.

[75] Drug resistance in fungi − an emerging problem (Online). Available from <http://www.searo.who.int/publications/journals/regional_health_forum/media/2011/V15n1/rhfv15n1p97.pdf>. (accessed 15.09.15).

[76] P.K. Mukherjee, J. Chandra, D.M. Kuhn, M.A. Ghannoum, Mechanism of fluconazole resistance in Candida albicans biofilms: Phase-specific role of efflux pumps and membrane sterols, Infect. Immun. 71 (8) (2003) 4333−4340.

[77] M. Sanguinetti, B. Posteraro, C. Lass-Flörl, Antifungal drug resistance among Candida species: mechanisms and clinical impact, Mycoses. 58 (2015) 2−13.

[78] I.S. Correia, The yeast ABC transporter Pdr18 (ORF YNR070w) controls plasma, Biochem. J. 440 (2011) 195−202.

[79] R.D. Cannon, E. Lamping, A.R. Holmes, K. Niimi, P.V. Baret, M.V. Keniya, et al., Efflux-mediated antifungal drug resistance, Clin. Microbiol. Rev. 22 (2) (2009) 291−321.

[80] R.E. Lewis, P. Viale, D.P. Kontoyiannis, The potential impact of antifungal drug resistance mechanisms on the host immune response to Candida, Virulence 3 (4) (2012) 368−376.

[81] D.M. Dixon, T.J. Walsh, Medical Microbiology, in: S. Baron (Ed.), Chapter 26: Antifungal Agents, 4th edition, Galveston (TX): University of Texas Medical Branch, Galveston, 1996.

[82] K. Nakamura, M. Niimi, K. Niimi, R. Ann, J.E. Yates, A. Decottignies, et al., Functional expression of Candida albicans drug efflux pump Cdr1p in a Saccharomyces cerevisiae strain deficient in membrane transporters, Antimicrob. Agents. Chemother. 45 (2001) 3366−3374.

[83] P. Vandeputte, S. Ferrari, A.T. Coste, Antifungal resistance and new strategies to control fungal infections, Int. J. Microbiol. 2012 (2012).

[84] S. Shukla, T. Prasad, and R. Prasad, Molecular Mechanism of Antifungal Resistance, pp. 1−15, 2006.

[85] S. Dogra, S. Krishnamurthy, V. Gupta, B.L. Dixit, Asymmetric distribution of phosphatidylethanolamine in C. albicans: possible mediation by CDR1, a multidrug transporter belonging to ATP binding cassette (ABC) superfamily, Yeast 121 (1999) 111−121.

[86] J. Morschhäuser, The genetic basis of fluconazole resistance development in Candida albicans, Biochim. Biophys. Acta - Mol. Basis Dis. 1587 (2−3) (2002) 240−248.

[87] M. Gaur, N. Puri, R. Manoharlal, V. Rai, G. Mukhopadhayay, D. Choudhury, et al., MFS transportome of the human pathogenic yeast Candida albicans, BMC Genomics 9 (2008) 579.

[88] S.S. Pao, I.T. Paulsen, M.H. Saier, Major facilitator superfamily, Microbiol. Mol. Biol. Rev. 62 (1) (1998) 1−34.

[89] R. Prasad, M.K. Rawal, Efflux pump proteins in antifungal resistance, Front. Pharmacol. 5 (2014) 1−13.

[90] S.S. Krishnamurthy, R. Prasad, Membrane fluidity affects functions of Cdr1p, a multidrug ABC transporter of Candida albicans, FEMS Microbiol. Lett. 173 (2) (1999) 475−481.

[91] A.H. Shah, A. Singh, S. Dhamgaye, N. Chauhan, P. Vandeputte, K.J. Suneetha, et al., Novel role of a family of major facilitator transporters in biofilm development and virulence of Candida albicans, Biochem. J. 460 (2) (2014) 223−235.

[92] C. Costa, P.J. Dias, I. Sá-Correia, M.C. Teixeira, MFS multidrug transporters in pathogenic fungi: do they have real clinical impact? Front. Physiol. 5 (2014) 197.

[93] P. Borst, Transporters as drug carriers: structure, function, substrates, Clin. Pharmacol. Ther. 88 (5) (2010) 578−579.

[94] J. Smith, D. Andes, Pharmacokinetics of antifungal drugs; implications for drug selection, Infect. Med. 23 (2006) 328−333.

[95] X. Li, Y. Hou, L. Yue, S. Liu, J. Du, S. Sun, Potential targets for antifungal drug discovery based on growth and virulence in *Candida albicans*, Antimicrob. Agents Chemother. 59 (2015) 5885−5891.

[96] H.I. Zgurskaya, H. Nikaido, Multidrug resistance mechanisms: drug efflux across two membranes, Mol. Microbiol. 37 (2000) 219−225.

[97] J. Higgins, E. Pinjon, H.N. Oltean, T.C. White, S.L. Kelly, Triclosan-mediated antagonism of fluconazole activity against Candida albicans requires EFG1, J. Dent. Res. 91 (2012) 65−70.

[98] T. Prasad, S. Hameed, R. Manoharlal, S. Biswas, C.K. Mukhopadhyay, S.K. Goswami, et al., Morphogenic regulator EFG1 affects the drug susceptibilities of pathogenic Candida albicans, FEMS Yeast Res. 10 (5) (2010) 587−596.

[99] D. Sanglard, A. Coste, S. Ferrari, Antifungal drug resistance mechanisms in fungal pathogens from the perspective of transcriptional gene regulation, FEMS Yeast Res. 9 (7) (2009) 1029−1050.

[100] N.A. Gaur, R. Manoharlal, P. Saini, T. Prasad, G. Mukhopadhyay, M. Hoefer, et al., Expression of the CDR1 efflux pump in clinical Candida albicans isolates is controlled by a negative regulatory element, Biochem. Biophys. Res. Commun. 332 (1) (2005) 206−214.

[101] In vivo and in vitro testing − Free in vivo and in vitro testing information|Encyclopedia. com: Find in vivo and in vitro testing research (Online). Available from <http://www. encyclopedia.com/doc/1G2-3404000449.html>. (accessed 01.03.16).

[102] M.R. Fielden, K.L. Kolaja, The role of early in vivo toxicity testing in drug discovery toxicology, Expert Opin. Drug Saf. 7 (2) (2008) 107−110.

[103] S.S.W. Wong, R.Y.T. Kao, K.Y. Yuen, Y. Wang, D. Yang, L.P. Samaranayake, et al., In vitro and in vivo activity of a novel antifungal small molecule against Candida infections, PLoS One 9 (1) (2014) e85836.

[104] D. Andes, In vivo pharmacodynamics of antifungal drugs in treatment of candidiasis, Antimicrob. Agents Chemother. 47 (2003) 1179−1186.

[105] M.B. Kurtz, G. Abruzzo, A. Flattery, K. Bartizal, J.A. Marrinan, W. Li, et al., Characterization of echinocandin-resistant mutants of Candida albicans: genetic, biochemical, and virulence studies, Infect. Immun. 64 (8) (1996) 3244−3251.

[106] A. Biernasiuk, A. Malm, S. Kiciak, K. Tomasiewicz, Susceptibility to antifungal drugs of Candida albicans isolated from upper respiratory tract of patients with chronic hepatitis C, J. Pre. Clin. Res. 7 (2013) 111−113.

[107] M. Iñigo, J. Pemán, J.L. Del Pozo, Antifungal activity against Candida biofilms, Int. J. Artif. Organs 35 (10) (2012) 780−791.

[108] Y.H. Samaranayake, J. Ye, J.Y.Y. Yau, B.P.K. Cheung, L.P. Samaranayake, "In vitro method to study antifungal perfusion in Candida biofilms", J. Clin. Microbiol. 43 (2) (2005) 818−825.

[109] H.H. Lara, D.G. Romero-Urbina, C. Pierce, J.L. Lopez-Ribot, M.J. Arellano-Jiménez, M. Jose-Yacaman, Effect of silver nanoparticles on Candida albicans biofilms: an ultrastructural study, J. Nanobiotechnology 13 (2015) 91.

[110] D. Andes, D.J. Diekema, M.A. Pfaller, J. Bohrmuller, K. Marchillo, A. Lepak, "In vivo comparison of the pharmacodynamic targets for echinocandin drugs against Candida species", Antimicrob. Agents Chemother. 54 (6) (2010) 2497–2506.

[111] D.R. Andes, D.J. Diekema, M.A. Pfaller, K. Marchillo, J. Bohrmueller, In vivo pharmacodynamic target investigation for micafungin against Candida albicans and C. glabrata in a neutropenic murine candidiasis model, Antimicrob. Agents Chemother. 52 (no. 10) (2008) 3497–3503.

[112] D. Andes, D.J. Diekema, M.A. Pfaller, R.A. Prince, K. Marchillo, J. Ashbeck, et al., In vivo pharmacodynamic characterization of anidulafungin in a neutropenic murine candidiasis model, Antimicrob. Agents Chemother. 52 (no. 2) (2008) 539–550.

[113] Bioavailability – definition and examples (Online). Available from <http://pharma.about.com/od/B/g/Bioavailability.htm>. (accessed 01.03.16).

[114] J.A. Barone, B.L. Moskovitz, J. Guarnieri, A.E. Hassell, J.L. Colaizzi, R.H. Bierman, et al., Enhanced bioavailability of itraconazole in hydroxypropylbeta-cyclodextrin solution versus capsules in healthy volunteers", Antimicrob. Agents Chemother. 42 (7) (1998) 1862–1865.

[115] F.M. Uckun, S. Qazi, S. Pendergrass, E. Lisowski, B. Waurzyniak, C.-L. Chen, et al., In vivo toxicity, pharmacokinetics, and anti-human immunodeficiency virus activity of stavudine-5'-(p-bromophenyl methoxyalaninyl phosphate) (stampidine) in mice, Antimicrob. Agents Chemother. 46 (11) (2002) 3428–3436.

[116] D. Andes, Antifungal pharmacokinetics and pharmacodynamics: understanding the implications for antifungal drug resistance, Drug Resist. Updat. 7 (2004) 185–194.

[117] E.S.D. Ashley, R. Lewis, J.S. Lewis, C. Martin, D. Andes, Pharmacology of systemic antifungal agents, Clin. Infect. Dis. 43 (Suppl. 1) (2006) S28–S39.

[118] P.W. Sylvester, Drug design and discovery, Drug Des. Discov. Methods Mol. Biol. 716 (2011) 157–168.

[119] G.M. Morris, M. Lim-Wilby, Molecular docking, Methods Mol. Biol. 443 (2008) 365–382.

[120] S.B. Bari, N.G. Haswani, Design, synthesis and molecular docking study of substituted N-aminocarbonyl arylvinylbenzamides, Int. J. Pharma. Res. Rev. 2 (2013) 6–19.

[121] M. Baginski, K. Sternal, J. Czub, E. Borowski, Molecular modelling of membrane activity of amphotericin B, a polyene macrolide antifungal antibiotic, Acta Biochim. Pol. 52 (2005) 655–658.

[122] I.M. Kapetanovic, Computer-aided drug discovery and development (CADDD): in silico-chemico-biological approach, Chem. Biol. Interact. 171 (2) (2008) 165–176.

[123] D.B. Kitchen, H. Decornez, J.R. Furr, J. Bajorath, "Docking and scoring in virtual screening for drug discovery: methods and applications", Nat. Rev. Drug Discov. 3 (11) (2004) 935–949.

[124] G. Sliwoski, S. Kothiwale, J. Meiler, E.W. Lowe, "Computational methods in drug discovery", Pharmacol. Rev. 66 (1) (2014) 334–395.

Printed in the United States
By Bookmasters